CREATURES
of
DARKNESS

CREATURES
of
DARKNESS

DANI ROBERTSON

100 of the Planet's
Weird and Wonderful Animals
That Come Out at Night

Harper North

HarperNorth
Windmill Green
24 Mount Street
Manchester M2 3NX

A division of
HarperCollins*Publishers*
1 London Bridge Street
London SE1 9GF

www.harpercollins.co.uk

HarperCollins*Publishers*
Macken House, 39/40 Mayor Street Upper
Dublin 1, D01 C9W8, Ireland

First published by HarperCollins*Publishers* 2025

1 3 5 7 9 10 8 6 4 2

© Dani Robertson 2025

Dani Robertson asserts the moral right to be identified as the author of this work

A catalogue record of this book is available from the British Library

HB ISBN 978-0-00-872883-0

Printed and bound in the UK using 100% renewable electricity at CPI Group (UK) Ltd

MIX
Paper | Supporting
responsible forestry
FSC
www.fsc.org
FSC™ C007454

This book contains FSC™ certified paper and other controlled sources to ensure responsible forest management.

For more information visit: www.harpercollins.co.uk/green

For all my fellow Creatures of Darkness, and to Ryan, who is simply a most marvellous creature.

Contents

Diurnal
Used to describe species that are predominantly active during daylight hours. Taken from the Latin *diurnus* (daily). Meaning 'belonging to the day'.

Crepuscular
In zoology, this refers to species that are active at dawn or dusk. Taken from the Latin *crepusculum* (twilight) and *creper* (uncertain, dusky, dark).

Nocturnal
Used to describe species active during hours of darkness. Taken from the Latin *nocturnus* (belonging to the night).

Cathemeral
Refers to animals that are not truly any of the above, changing their periods of activity to suit their environment. Taken from the Greek *kata* (through) and *hemera* (day) with day here meaning a whole night and day.

Most of us assume that living creatures prefer to be active in the light and sunlit hours of the daytime. When the Sun rises and the birds sing, we look to the beautiful colours of a dawning day and feel life bursting from all around us. However, when the Sun starts to set and the stars come out to shine, that's when the planet really comes alive. This, is the rhythm of the night.

This book will take you on a global journey, gaining you special access to the planet's most exclusive nightclub – our nocturnal world.

These pages will peel back the deepest, densest rainforest foliage and unmask the true rumble in the jungle. Stars will sparkle over wide-open savannahs and deserts at dusk, before taking a tumble through the humble rural countryside, peeking at the astonishing world that goes unseen every night. We'll hold our breath before plunging to the darkest depths of the known ocean, where we will traverse canyons and trenches that have never felt the warmth of our sun to discover life that lives in an atramental eternity. Subterranean secrets will be revealed as we journey underground, opening eyes and minds to how life thrives in the very darkness humans try their best to erase.

Whole new worlds and ways of living will be revealed to you as we take a look beneath the cover of darkness. You will be introduced to the weirdest (kangaroos that live in trees) and most wonderful (tiny armadillos) creatures that have taken to the dark and made it their home. A world where deception is deployed in the most marvellous ways. You will meet stargazing beetles, perplexing parrots, and magnificent mammals, all of whom live their life by the light of the stars and the Moon.

As humans, we try to make life on this planet exist in a way and within a pattern that fits our own experience of the night and day, but of course, time is a concept constructed by humans to bring sense and structure into the chaos of our lives. We have long forgotten the pull of the Moon and the lure of a starlit night on the human mind, as we have created our little artificial suns that we hang in our homes and along our streets.

The same cannot be said for our wildlife, most of which is governed not by the 24 hours in a human day, but by the lunar cycle, the rhythm of the Sun's rising and falling and the turning of the Milky Way arching up over us in the heavens.

Our installation of artificial suns and moons all over the place, is causing chaos and confusion in the animal world. Darkness is a habitat that has been neglected, forgotten as soon as we first turned on the lights. It is one of the most endangered habitats in the world, and the threat grows each year as light pollution threatens to drown even the deepest, darkest parts of the planet.

A world so secret, many of these species have been seen by just one or two humans, sometimes centuries apart. We know so little about the planet's night life, that we simply don't understand why some of the animals behave the way they do, nor if they truly fit into the categories of nocturnal or diurnal.

Every day (well, night) zoologists, conservationists and ecologists are adding to this list, as we clamour to learn as much as we can about this hidden world. The clock is against us, as so many of the species in these pages are being pushed to the very brink of existence. Each species' profile will show how it is listed on the International Union for Conservation of Nature (IUCN) or other conservation lists. Many of them will be lost to the planet and lost to science before we can truly understand their role in the darkness, and we are all much poorer for each species we lose. At the back of this book, you will find a list of the organisations that are working around the clock to help these wonderful animals see another night.

All these animals need darkness to survive. Light pollution is having a devastating impact on our wildlife, with even a single bulb having the potential to destroy a vital habitat and completely deconstruct an ecosystem, in an instant. The power lies, quite literally, at our fingertips, to have the ability to do something amazing for life on the planet and to stop light pollution.

Embrace the dark side and release the natural night into your world by flicking that switch. Return the dark to the night and restore the skies by taking out that wall of light. Let that starlight stream in, like a great cosmic river of wonder. By reinstating darkness we are doing vital habitat restoration works, sometimes from the comfort of our own homes. You will be helping the wildlife you're about to meet among these pages go about its night life the way it is intended. With natural order restored you will be giving it a fighting chance to thrive. The darkness needs us. And we need the darkness.

European Hedgehog

European Hedgehog

Erinaceus europaeus

Draenog • Furze-pig • Hotchi-witchi

Nocturnal

Location	Average Lifespan	Average Size	Average Weight	IUCN Red List Status
Western Europe	2–3 years	15–30cm long	2kg	Vulnerable to extinction (UK)

In Britain long ago, the turbulent Tudors persecuted the hedgehog for it was believed they stole milk from cows, like a thorny thief in the night. Those producing a dead 'urchin' (not a child, but an Old English colloquialism for hedgehogs) were rewarded with a groat, equating to four pennies in today's money, which was half the daily wage of a skilled labourer back in the 1480s. This sadly led to widespread slaughter of the harmless hog for over 200 years. We now know hedgehogs are lactose intolerant, so if you do feed them, make sure it is water, not milk.

The humble hedgehog, a visitor to our gardens, parks and country lanes. An unmistakable and iconic creature that most westerners have stumbled upon one time or another when out on an autumnal walk. You'll find this prickled prince of the hedgerow snuffling about in a garden leaf pile or shuffling along the dark edges of parkland, looking for a warm and comfortable place to make their home for winter hibernation.

They are omnivores and though their button black eyes are twinkly bright, their eyesight is poor. They depend on their excellent noses and petal shaped ears to hunt at night for insects such as slugs, worms and even amphibians. Get too close and you'll see that elegant snout disappear, as the hedgehog's party trick sees it rapidly turn into a barbed ball, protecting it from predators and putting those 7,000 spikes to excellent use as a homegrown fortress.

Our bristly little friend of the hedge is truly iconic of the night. If you should like to help them survive into the next century, please take care when gardening. Many fall foul to the busy blades of strimmers and mowers, as they lay unseen in garden edges and leaf piles or face painful deaths after ingesting garden chemicals such as slug repellent.

Our nights would not be the same without these little hogs.

Barn Owl

Tyto alba

Banshee • Tylluan Wen • Night Owl • Scréachóg Reilige

Nocturnal

Location	Average Lifespan	Average Weight	UK Red List for Birds	European Birds of Conservation Concern List
Widespread globally	4 years (20 in captivity)	430–620g	Least concern	Amber

As the Sun sinks into golden hour, an apparition gazes through a weathered window of a ramshackle barn. A perfect white heart punctured by two bottomless wells of deepest, darkest black, stare into the evening's vanishing light, watching all and missing nothing. As twilight takes hold, the spectre spreads its wings, gliding out into the night, absorbing all the details that are shrouded to human eyes by shadows.

Silently she sails over her domain, searching for a poor soul to feed her brood. Razor-sharp talons descend from the sky, much to the surprise of the unfortunate rat. Her wings are designed for completely silent flight, her eyes, two powerful pools of darkness, give her excellent night vision, while her ears are the most sensitive of nearly any animal ever tested on the planet. Barn owls are designed to be deathly in the darkness. A screeching and hissing call has found her many names in many tongues, such as *Scréachóg Reilige*, which translates to 'the graveyard screamer'.

Barn owls were once abundant, a farmer's friend that cleared the yard of scurrying vermin and seen hunting along miles of hedgerows. But our lives have changed and shaped a new landscape, one that doesn't have room for hedgerows and tumbledown barns. It is mainly this intensification of agriculture that has seen 118,000 miles of hedgerows lost since 1950. The humble hedgerow may not look like much, but it is home to over 1,500 invertebrate species. They also create food for the barn owl. With our clearing up of the countryside, the tumbledown barns are polished into respectable renovations, removing any potential room to roost. The barn owl's home, once sheltered and shadowed, is now dressed in stone and finished with a flourish in the form of a floodlight, demolishing the dark and hallowed hunting halls once and for all.

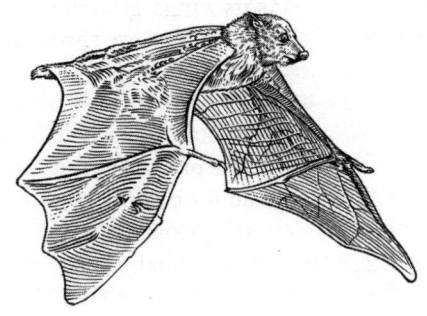

Spectacled Flying Fox

Pteropus conspicillatus

Giraman • Bilihny • Manu-wudhaga

Nocturnal

Location	Average Lifespan	Average Wingspan	Average Weight	IUCN Red List Status
Northeast Australia, New Guinea, and some offshore islands	12–15 years (30 in captivity)	up to 1m	up to 1kg	Endangered

As dusk settles to cool down a humid day, spectacled flying foxes take to the skies, their giant wings silhouetted against a lavender sky as they commute to their night shift. A type of Megabat, these giant fruit bats are creators of the rainforests. Feasting on fruits, nectar and pollen a single bat can cover nearly 70 miles and disperse 60,000 seeds in a single night – making them vital for pollination and seed dispersal, and a crucial recruit in our fight against climate change.

During the day they can be found in their tropical canopy 'camps' where this social species hang out, often in their thousands, toes skyward, wings tightly wrapped like a batty burrito. They are highly intelligent, with doleful brown eyes bespectacled by frames of golden fur. Not just there to make them look clever, their eyesight is excellent for night-time foraging, so good in fact that they go without the echolocation relied upon by so many other of their bat cousins.

They do so much to help biodiversity and in times gone by they would leave a trail behind them of blossoming new plant life. Now they are threatened by habitat clearance and human impacts, most shockingly the result of a rapidly changing climate. In November 2018, a deadly heatwave roasted Australia with temperatures exceeding 42°C, the highest ever recorded temperature in Cairns. This had a devastating consequence on these furry foresters, who could not cope with the extreme heat, dying from heat fatigue in their thousands.

Over just 2 days, November 26 and 27, over 23,000 spectacled flying foxes died, a third of Australia's entire population. Their situation is incredibly precarious, just one more extreme heat event like November 2018 could see this species become extinct, disappearing from the night skies and ceasing their rainforest regeneration in an instant.

European Nightjar

Caprimulgus europaeus

Goatsucker • Troellwr Mawr • The Fern Owl • Purrin Bird • Corpse Bird • Nattraven

Nocturnal

Location	Average Lifespan	Average Size	Average Weight	Red List for Birds
Migratory, spanning much of North and Central Europe and North Central Asia	2–3 years	25–28cm	50–95g	Amber (UK)

The king of camouflage, this enigmatic bird species is shrouded in mystery and folklore that goes deeper than the darkness of a moonless night. They are skilled in the dark arts, shrouding themselves in a cloak of invisibility by daylight. Their feathers appear exquisitely crafted from bark, a bird intrinsic to the darkness of the woodland floor where they blend seamlessly into the scene, shaped more like a squat toad than a hawk. Their eyes, two oracular orbs fill most of its face, ending in a deceivingly small, pointed and peculiarly whiskered beak (known as rictal bristles). When hunting, the lower jaw opens so wide it appears almost unhinged, this shockingly wide mouth (or 'gape') allows it to feast effortlessly on large insects without the need to take its prey apart.

Cloaked by nightfall they erupt from their lowly perch to shatter the sky with their churring call, wings outstretched akin to a kestrel. Humans are far more likely to hear, rather than see this covert creature. Their 'clapping' beats adding a bassline to the rhythm of the night as the males hit their wing tips together in the quest to attract a mate. 'Nattraven', night raven, has accompanied Odin on the Wild Hunt, and been accused of aligning with malignant forces to steal milk from goats and the spirits of children. In reality, this mysterious creature feasts throughout the night, on moths, beetles and flies, utilising its wide mouth, silent flight and darkness to become the ultimate nocturnal hawker.

On warm and breathless evenings in early June, take yourself to the edges of a woodland and allow the colours of the day to drain from the sky. Sit and listen, and you may be rewarded with the futuristic and fantastical sounds of this furtive feathered friend of the night.

Eurasian Otter

Lutra lutra

Dyfrgi • Dobharchú • Dratsie

Nocturnal and Crepuscular

Location	Average Lifespan	Average Size	Average Weight	IUCN Red List Status
Ireland to China, Southeast Asia	5–10 years	60–80cm long (without tail)	6–8kg	Near threatened

The river's waters hold the darkness of the skies as it snakes through the reeds, bubbles rising to the surface before a glimmering serpentine breaks the boundary of water and air. First, a pair of flaring nostrils adorned with slender, twitching whiskers, exhale a huff of air, sprinkling droplets of river skyward followed closely by two keen and knowing eyes that peer from the rivery meniscus. The noble soul of the river and warden of the water's edge, the otter transcends land and liquid, shapeshifting as she goes. Slickened fur gleaming, viscous like liquid jet, slides lithely between the riverbed and bank. In a blink she pirouettes into a dive leaving only a memory of bubbles behind.

Their world is not only underwater, but underground. Otters retreat to a dark and cosy holt to sleep after a busy night hunting trout, eel and mussels. Darkness offers safety and repose from the threats of daylight. Their whiskers like fibre optics and primed to find prey in the water's umbra effortlessly, taken in their powerful webbed paws and feasted on either on the river's banks or taken back to the holt to share. They may be known as water dog, but a happy otter purrs like a cat.

Humans have long admired the otter, with fabled Otter Kings in Scotland whose otter contingents could grant wishes and their skins believed to give protection from drowning. They even have a Patron Saint in St Cuthbert, who offered them his protection as thanks to the creatures for warming his feet after his nightly prayer vigils, done standing waist deep in the boreal and brumal waters of the North Sea.

Binturong

Arctictis binturong

Bearcat • Maturun • Malay Civet Cat

Nocturnal

Location	Average Lifespan	Average Size	Average Weight	IUCN Red List Status
South and Southeast Asia	4 years (20 in captivity)	60–96cm long (without tail)	10–14kg	Vulnerable

A creature as ancient and mysterious as the night itself is the gloriously bizarre binturong. Its name derives from the ancient indigenous languages of Borneo and Sumatra, and the earliest form we know of this name is *maturun* meaning 'one who descends'. This perfectly describes this feline-esque forest furitive's behaviour; instead of leaping from branch to branch, they usually leave the canopy and walk across the forest floor to climb the next tree. By day this weary-looking creature stays in the forest canopy, snoozing the day away. Curled up like a cat, its wiry black and silver flecked fur glistens in the afternoon sun while its super long, prehensile tail acts as anchor on a tree branch. Frequently confused to be some sort of cat, the binturong adds to this mistaken identity, as they purr when pleased and can even chuckle.

At night it must descend to find food, actively seeking out fruits, berries and small animals for a midnight feast. Speaking of snacks, they have a surprising scent, smelling like hot, buttered popcorn. They act as important seed spreaders, their nightly activities crucially continuing the circles of life on Earth, proving that night is as important as day when it comes to our wonderful world of wildlife.

Malu the binturong made headlines in the UK in 2024. The 'shy yet friendly creature' managed to evade his keepers at Dartmoor Zoo, taking himself off for a few days to explore the Devonshire countryside before being found, safe and sound, in a rustic woodshed alongside an allotment which he was using for food. His holiday was soon over and he was returned to the safety of his enclosure at the zoo.

Giant Anteater

Myrmecophaga tridactyla

Antbear • Tamanduá-bandeira

Nocturnal and Diurnal

Location	Average Lifespan	Average Size	Average Weight	IUCN Red List Status
Central and Southern America	15 years (25 in captivity)	1.8-2.4m long	33–50kg	Vulnerable

Many of our most curious creatures are most active at night. This one is curious indeed. It moves like a bear walking on elongated stilts, draped in flowing fur. A long, tubular, elephantine snout contains an even longer tongue, which it can extend up to 50 centimetres. In fact, it has the longest tongue of any land mammal on the planet. It has two huge claws on each front paw, perfectly equipped to tear open an ant hill. It keeps its claws sharpened by walking on its knuckles to protect them. With no teeth, it slurps upwards of 30,000 ants a day. It is so well adapted that it can even eat the aggressive fire ants. Its sense of smell is forty times more sensitive than humans, meaning darkness is no barrier to finding food. It is the largest of the anteater species alive today.

Its name in Portuguese, *tamanduá-bandeira,* means 'flag anteater' and refers to its fabulously flamboyant tail, like a bristly battle standard streaming behind it. During the hot days this tail can be used as a sunshade while the anteater sleeps. They live in the grasslands, savannahs and rainforests of the Americas.

A prevalent issue with being a nocturnal creature is it almost always arouses human suspicion. In one form of the Brazilian folklore story, the Capelobo is a monster with the head of a giant anteater and body of a human that hunts humans for their blood. A little like a tubular-snouted Count Dracula.

Sadly, these comical looking creatures are the most threatened mammal in Central America. The main threat this gentle giant faces is habitat loss due to human agricultural practices and fires. They are also hunted as pests. But all is not lost, conservation efforts are taking place to safeguard this solitary creature of the night.

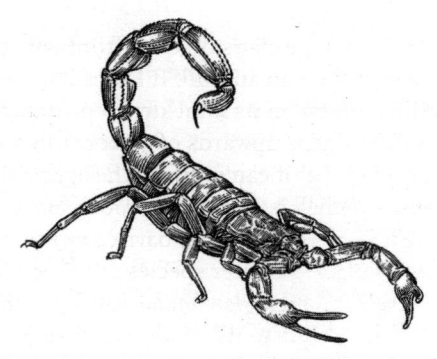

Deathstalker Scorpion

Leiurus quinquestriatus

Palestine Yellow Scorpion • Omdurman Scorpion • Naqab Desert Scorpion

Nocturnal

Location	Average Lifespan	Average Size	Average Weight	IUCN Red List Status
North Africa, Middle East, Sahara Desert	Unknown in wild	6–10cm long (including tail)	1–2.5g	Unevaluated

Somewhere deep in the burning desert waits a swift and effective hunter. Among the vast expanses of sand, the deathstalker scorpion patiently waits, so still it could be carved from the sand itself. It is not a creature to be messed with, semaphoring this to the world with its bright, acrid-yellow colouring and defiantly curled tail, complete with razor-sharp stinger, primed and ready to make itself known to anyone who crosses its path. It is one of the most venomous scorpions in the world; its venom being a powerful mixture of neurotoxins that deliver a hefty punch.

Almost all scorpions are nocturnal. To deal with the heat of their furnace-like habitats, they will burrow during the day, to wait out the heat, before emerging into complete darkness to hunt. Standing statuesquely as their prey approaches, the hairs on the scorpion's body vibrate, and the hunter takes its shot, piercing its victim's skin and envenomating them.

Scorpions also have a curious ability to glow in the dark. UV lights can reveal a whole new world lying there in the darkness, but scorpions really glow. Even fossilised ones! Science is still trying to find the reason why, but theories include the UV glow acting like sunscreen, helping scorpions recognise each other or even using their glow to dazzle their prey making it easier to hunt.

Of course, a scorpion resides in the heavens, Scorpius, the giant scorpion sent to Earth by the Greek goddess, Gaia, great mother of all creation, to defend the Earth's animals from the hunter Orion who claimed to want to kill all the animals of the world. Scorpio battled Orion and as a reward for his bravery, Gaia placed the mighty scorpion in the stars, where he eternally chases Orion throughout the night sky.

Raccoon

Procyon lotor

Mapachitli • Aroughcun

Nocturnal

Location	Average Lifespan	Average Size	Average Weight	IUCN Red List Status
North America	3–5 years	90cm long	4–9kg	Least concern

This comically masked creature, like the bandit Zorro, with black markings across its eyes, is one of the true masters of the nocturnal environment. The mask that gives them so much character even helps them see at night, the black fur absorbing light, reducing glare to help them find the next meal with ease. They have no fear of the dark, nor do they shy from the light. These valiant pioneers have made themselves right at home in our urban cities, living among humans and outwitting us at every turn. Their curiosity and clever minds have made them a true emblem of the night. Their ability to find food among human rubbish means these mischief-makers are often getting themselves into trouble with humans. Armed with dexterous front paws, they can unscrew jars and twist open door handles, sometimes leading to property being damaged and general nuisance-causing, such as sending rubbish bins clattering into the middle of the night. Of course, this is a manmade problem that raccoons are adapting to with the resources available to them.

The February Full Moon is given the moniker of the 'Raccoon Moon' by the Lakota tribe. Many Indigenous American peoples have folklore attributing to *aroughcun*, the Powhatan name that the name 'Raccoon' comes from. These creatures are respected for their intelligence and dexterity, but also their playful nature.

So playful in fact, they easily capture the heart of humans who have any interaction with this species, like 'Raquinho', the Raccoon who found himself an unlikely sports star as he invaded the pitch of a Major League Soccer match between Philadelphia Union and New York City FC. He soon had the crowd eating from his dexterous paws, showing a finesse for fancy footwork and impressive evasive skills. The crowd and match commentators went wild, rooting for Raquinho as he dodged his would-be captors. One commentator even suggested signing the skilful upstart, affectionately given his moniker in reference to the Brazilian football legend Ronaldinho.

Virginia Opossum

Didelphis virginiana

Tacuachi

Nocturnal

Location	Average Lifespan	Average Size	Average Weight	IUCN Red List Status
North America	1–2 years (4+ in captivity)	36–48cm long (excluding tail)	1.9–2.8kg	Least concern

This solitary soul slinks around through the night, keeping itself to itself, almost as if it knows how misunderstood by humans it is. Named for its white face, it's a furry little fellow and is well adapted for nocturnal ventures, whose creamy white fur glows softly under starlight, merging into a mottled moon grey around its ears and the rest of its body. It has a long, hairless, prehensile tail that it puts to good use when climbing.

It is the only marsupial native to North America and also has the most teeth of any mammal, with a set of fifty. Although it can look ferocious when baring its impressive set of teeth, when threatened the opossum simply keels over, its life seemingly extinguished. But worry not, this is a survival tactic known as 'playing possum'. They are not harmful to humans, being timid creatures that try to stay out of our way. They spend their days dreaming away in a den, usually inside a hollow tree or abandoned burrow.

Doting mothers, female opossums have large litters of offspring. These jellybean sized Joeys live in their mother's pouch until they outgrow it. Then the mum must carry her offspring on her back, like multiple furry backpacks hitching a lift until they're old enough to survive alone.

In ancient Mesoamerica the humble opossum was seen as a bit of a Robin Hood character, who 'stole' fire from the gods to bring it to humans. The story goes that the opossum once had a gloriously thick and bushy tail, but it used this to catch the fire, sacrificing his furry tail to bring fire to human beings.

Oppossum have presidential approval, with America's twenty-third president, Benjamin Harrison, being a big fan of the opossum, bringing two to live at the White House during his presidency — Mr Protection and Mr Reciprocity. President Herbert Hoover also had a pet opossum, named Billy, who the family adopted from the wild.

European Badger

Meles meles

Moch Ddaer • Broc • Bawson • Brochlach • Stiall-chù • Borsuk • Tejon

Nocturnal

Location	Average Lifespan	Average Size	Average Weight	IUCN Red List Status
Europe, West and Central Asia	3–5 years	90cm long (excluding tail)	10–12kg	Least concern

Clad in black-and-white-striped pyjamas, badgers personify the night as they bumble along the forest edges. Stout legs keep their barrel-like bodies close to the floor as they hunt insects, small animals and tubers, sniffed out by their elegant striped snout, and prised from the earth by their powerful, clawed paws. You may just see the unmistakable rump, adorned by a small puff of tail, disappear through a hedge. A mustelid, they belong to the same family as weasels, stoats and otters and live in social family groups of six or more adults. They're so social that they have been known to coexist with another species, red foxes.

The full and luminous October Moon was once known in Scots Gaelic as *'Gealach Bhuidhe nam Broc'*, the yellow Moon of the badgers. This name alludes to the behaviour of the fastidious badger as winter approaches, cleaning out its sett and tumbling grass by the light of the Moon to dry it, in order to line its underground snug with a luxurious bed of hay ahead of its peaceful slumber through the harsh winter weather, hopefully safe from interruption at the hands of man. They don't hibernate, but go into a state known as 'winter lethargy', like many humans I know.

In Irish mythology, the King of Tara, Tadg, is said to have kinsmen who could shapeshift into the form of badgers. They are revered across many countries and appear across cultures, such as German folklore in which the badger is a homebody and devotee to his family, wanting nothing but a quiet and peaceful life above all else.

North Island Brown Kiwi

Apteryx mantelli

Te manu huna a Tāne

Nocturnal

Location	Average Lifespan	Average Size	Average Weight	IUCN Red List Status
New Zealand	25–50 years	45-60cm	2.4–3.3kg	Vulnerable

This bird is the symbol of a nation, so treasured it has been blessed by the Māori God of the Forest, Tāne Mahuta. A selfless and brave bird, who according to legend sacrificed its wings and ability of flight by coming to Earth to protect the trees of the forest floors. These shy birds are loved ferociously by New Zealanders, who have adopted the affectionate moniker of 'Kiwis' for themselves. The kiwi is named for its call.

Living its life in the dense and tenebrous forest floor while New Zealand's starry skies twinkle above it, this cherished character resembles a ball of fur, its feathers looking more like long hairs than plumage. A long, protruding bill with two nostrils unusually perched at the very end of its beak – the only bird in the world to have so! – it is also adorned with cat-like whiskers, helping this curiosity clamber its way through the vegetation at night. Two glittering eyes sit like tiny gemstones, but offer little in the way of sight. Kiwis have adapted to the darkness by utilising smell, touch and hearing to delve through the darkness. They are a burrowing bird and can have over fifty burrows that they call home.

These enigmatic birds are at risk due to invasive and introduced species such as dogs, cats and mustelids, including stoats.

Such is its fame as a nocturnal crusader that Miami Zoo found themselves at the receiving end of a rightly wrathful nation, when New Zealanders found that a Kiwi bird was being used for 'meet and greets' under bright lights and flashing cameras. Within 24 hours of a video being posted on social media, Paora the Kiwi was returned to his rightful place – the darkness.

African Leopard

Panthera pardus

Ingwe • Nkwe • Chui

Nocturnal

Location	Average Lifespan	Average Size	Average Weight	IUCN Red List Status
Sub-Saharan Africa	10–12 years	60–70cm tall	Females 21-43kgs Males 31 – 72kgs	Vulnerable

High in the boughs of the Marula tree, a creature the colour of the savanna sun seeks shade. Precious gold glimmers, studded with rosettes of obsidian: a regal coat worthy of those who rule the savanna, rainforest, mountain and desert plain. Hidden from the clawing heat of the African sun, this noble nomad awaits the coming of its kingdom. Emerging only at the unveiling of dusk, revealing piercing emerald eyes that watch for the rising ceremony of the Moon.

With night vision seven times better than humans, the dark conceals the leopard in the long and swaying grasses as it stalks an antelope, lost from its herd at the edge of a busy watering hole. Springing from the shadows with power and precision, the antelope is swiftly taken, dragged back to the safety of the throne of the Marula tree. The leopard will banquet in its branches.

According to African tradition, it is not the lion, but the leopard, who is king. The leopard is seen as more skilful and wiser. They are believed to be protectors of humans against malevolent spirits. The Zulu call the leopard *ingwe* meaning 'the supreme ruler' reflecting this animal's symbolism for leaders and warriors who want to show they are courageous and noble. Zulus would once wear leopard-skin headbands to magically enhance their own human vision, as they believed that the leopard could see things so far away it would take hours to travel there on foot. Leopards are so strong they can drag a carcass three times their own weight 6 metres up into a tree.

African Savanna Elephant

Loxodonta africana

Indlovu • Tlou • Njovu

Diurnal and Nocturnal

Location	Average Lifespan	Average Size	Average Weight	IUCN Red List Status
Sub-Saharan, Central and Western Africa	60–70 years	3.4m tall	6 tons	Critically endangered

Reverent and wise, the African elephant is the largest land mammal in the world. Her wrinkled grey skin resembles the bark of the trees she feeds on, her ears the shape of the very continent she roams. Her giant form and thick skin belie the sensitive nature of this gentle beast. Her eyes are pools of amber honey, thick with the memories of a life on the run. Her long trunk can use its 150,000 muscles to lift a weight of 320 kilograms, yet is dexterous enough to pick a single, delicate flower from the ground unharmed. Her two tusks, the ivory hue of a summer's moon, have seen 90 per cent of her kind wiped from the planet in just a hundred years.

Elephants should need to sleep a lot, given their size, however, they are stupefying scientists, with the average elephant snoozing just 2 hours a day. They are active during daylight hours and during the night, when the cooler temperatures allow them to move more freely. The night-time offers protection from the Sun, but also poachers. Wise old elephant has adapted to move quickly and quietly through the night when around villages and humans. To elephants the night is a refuge from human eyes and harm.

For such a proud and noble creature to be reduced to moonlighting as a fugitive is a great shame upon humans. An old Kenyan tale tells the story of Elephant, Thunder and Humans existing on Earth together, but the conflict between the three saw Thunder leave the Earth. Elephant believed it could become equals with the Human and trusted them, but Human shot the Elephant. As Elephant lay dying it cried to Thunder to help it, but Thunder refused and said this was Elephant's punishment for trusting the Human. Now humans poach elephants to the brink of existence, and humans turn over its habitat for our own gain, leaving this mighty megafauna nowhere left to hide. Conservation efforts are hopeful we can halt the declines, but as they say, an elephant never forgets.

Red Kangaroo

Macropus rufus

Gambu Ganuurru • Kere Aherre • The Red Chief • Boomer

Diurnal and Nocturnal

Location	Average Lifespan	Average Size	Average Weight	IUCN Red List Status
Western and Central Australia	8 years (25 in captivity)	1.6m long (excluding tail)	47kg	Least concern

Kangaroos are mainly nocturnal as a species, preferring the cooler temperatures and protection from the Australian summer sun that the night offers, which allows them to go feasting on grasses and leaves without the prying eyes of predators. It's the water within these plants that make the red kangaroo a survival specialist, meeting their water needs through these plants when water is scarce. The females also have a unique ability to delay pregnancy when conditions are too dry to successfully reproduce. A group of kangaroos is fittingly known as a mob, given they pack quite the punch.

Synonymous with Australia, kangaroos have significant cultural importance for many First Nations People on the continent. An eastern bush kangaroo was once a bounding star of screen, but long before Skippy, the Boorong people of Victoria knew Purra, a kangaroo whose story is told in the stars. She is represented in the night sky by the star known to Western astronomy as Capella and pursued by two hunters, Yuree and Wanjel. Purra has even been aboard the International Space Station, and her story is reflected in the logo of the Australian Space Agency.

Back here on Earth, Kangaroos may be associated with the sunshine and heat of Australia, but in reality, these marsupials prefer the cool air of the night, using their powerful hind legs to bound vast distances, travelling in 'mobs' and covering up to 8 metres per jump! Their long tails are used for balance when feeding (and boxing). In some First Nation cultures the red kangaroo is a symbol of agility and balance, but also resilience given this species' adaptability to harsh changes in circumstance.

Koala

Phascolarctos cinereus

Gulamany • Kulla • Cullawine • Bangaroo

Nocturnal

Location	Average Lifespan	Average Size	Average Weight	IUCN Red List Status
Mainland Eastern and Southeastern Australia	12–15 years	70–90cm long (excluding tail)	6–8kg	Vulnerable

An iconic species that is perhaps best known for being sleepy. Snoozing soundly in the branches of a eucalyptus tree during the heat of the day, they look like small bears but are in fact marsupials, having more in common with kangaroos than actual bears. With fluffy grey coats and two crescent-shaped ears, tufts of white fur protrude haphazardly, giving them the somewhat dishevelled bedhead look. Not surprising given this sleepy head needs 18 hours of sleep a day! A large and leathery black nose makes up for their poor eyesight, as this king of the midnight feast needs to eat around 1 kilogram of leaves in a single night. Its large nose is used to sniff out the tastiest and most nutritious eucalyptus leaves.

The name 'koala' possibly originates from Dharug, an Aboriginal Australian language, meaning 'no drink'. This reflects the koala's behaviour in rarely drinking water, preferring to gain hydration from all those eucalyptus leaves, which most other species cannot tolerate due to the leaves being highly toxic. Primarily nocturnal, koalas prefer the shelter of the night to the scorching day.

Many eucalyptus trees seem to have adapted to nocturnal wildlife, producing more nectar and pollen at night, helping to support the night-time ecosystem and working together with other nocturnal treetop foragers, like flying foxes and many types of insects, to keep the chain of the ecosystem moving.

Koalas face many threats to their survival, with extreme heat events due to climate change ranking highly among the most common factors, alongside the conflicts that happen when human lives encroach on the homes of wildlife. Vehicle collisions are the second biggest threat to koalas, followed by attacks from domestic dogs. You can help koalas by driving more slowly and with greater caution, and by simply putting your pet pooch on a lead.

Attenborough's Long-beaked Echidna

Zaglossus attenboroughi

Spiny Anteater • Payangko

Nocturnal

Location	Average Lifespan	Average Size	Average Weight	IUCN Red List Status
Island of New Guinea	10–30 years	50–90cm long	5–10kg	Critically endangered

This elusive creature looks like it belongs within the pages of a fantasy novel, peculiar in its appearance and mysterious with its habits. They are rarely sighted in the wild, with 61 years passing between the last two scientific recordings of this fabulous beast, but I promise you that they do exist ... if you know exactly where to find them. A genuine curiosity to behold, its small, rotund body is covered in spines made from keratin, the same material that we find in our own fingernails. They look like a mash-up between a mole, hedgehog and porcupine. Its long snout, not too dissimilar to a bird's bill, helps it root around in the night for worms and insects, foraging on the forest floor. Even its home sounds like a fiction: they are known only to exist within the Cyclops Mountains. The last known sighting was in 1961, until on the very last day of a scientific expedition led by Oxford University and an Indonesian conservation group, Yappenda, in 2023. Upon inspecting the final memory card from trail cameras, scientists saw an image of this odd little fellow.

The name echidna was bestowed upon this species by European settlers in Australia who were well and truly stumped by how to label this 'new to science' critter. Not quite a mammal, not quite a reptile, they took the creature's namesake; the monstrous half-snake, half-woman in Greek myth. Of course, they could have just asked a local and saved themselves some time.

During the day it heads underground to its burrow, waiting for the protection of darkness to keep its life a mystery from the prying eyes of predators. Due to its feeding habits, it plays an important role in the local ecosystem, turning soil over and aerating the ground, improving the nutritional content and helping the local flora grow healthy and strong. Other than this, we know extremely little about this camera-shy soul.

Aye-aye

Daubentonia madagascariensis

Aiay

Nocturnal

Location	Average Lifespan	Average Size	Average Weight	IUCN Red List Status
Madagascar	10–23 years	70–90cm long	2–2.5kg	Endangered

This eternally startled-looking being is another misunderstood creature of the night. Wholly designed to live their lives at night, these lemurs look like electrocuted Gremlins. Coarse, black bushy hair covers its body and long bottlebrush tail, thinning out around its head and face, turning to wisps of fine-spun silver, which creates eyebrows that frame its most vital feature: two huge clementine eyes that pierce through the serried trunks of rainforest trees.

These ocularly gifted beasts combine their sight with two tremendous, bat-like ears that they use to sound out hollows in the wood of trees to identify exactly where their food (nests of insects) is likely hiding. Its hands – long, claw-like fingers with an extra-long central digit – are something even Tim Burton couldn't dream up. Its elongated digit helps to drag out, scoop up, and spoon in food from hard-to-reach places. Like a Swiss Army knife at the end of its arms, it taps branches at eight times a second to find its food, like a whimsical woodpecker. I refuse to believe this isn't where the famous extraterrestrial E.T. found inspiration for his hands.

During the day, they take to the heady heights of the rainforest canopy, weaving intricate nests into sumptuous spheres made of branches and leaves, sheltering themselves from the glare of the Sun and any predators lurking in the foliage.

They are unfortunate enough to bear the misnomer of being a bad omen. So feared, their name is thought to have come from the Malagasy phrase 'I don't know'– even saying its name out loud is enough to bring misfortune. This has led to a tradition of killing these peaceful and solitary souls, sadly very few now exist in the wild and they are listed as endangered.

Saltwater Crocodile

Crocodylus porosus

The saltie

Nocturnal

Location	Average Lifespan	Average Size	Average Weight	IUCN Red List Status
East coast India, Southeast Asia, Northern Australia and Micronesia	70 years	6.5m	1,000kg	Least concern

Crikey! The infamous catchphrase of Australia's beloved Crocodile Hunter, Steve Irwin. Before Irwin burst onto screens tumbling around with these huge reptiles, most people only knew crocodiles as the nemesis of Captain Hook and nothing more. Irwin educated the world on his favourite species, the saltwater crocodile, even naming his daughter after his beloved saltie, Bindi.

Not only is the saltwater crocodile the largest living reptile on the planet, it's also one of the oldest. They've been walking the Earth long before humans, stretching back 240 million years. We are talking the actual Jurassic Park. This species was once roaming the same land as the dinosaurs, and would have seen triceratops, velociraptors and even the mighty tyrannosaurus come and go. They somehow survived the extinction event that ended the age of the dinosaurs and have risen to the challenge of every era since.

A ferocious hunter, a large adult has a biting pressure of around 2 tonnes, to make sure you really feel each of those sixty-eight teeth in that crocodile's smile. They are largely nocturnal and sleep during the day, keeping one eye open while they warm up their cold blood in the sun. They are capable of unihemispheric sleep, an incredible evolutionary tactic that means they shut down half their brain at a time to rest, leaving the other half alert to any dangers (or their next meal).

Despite all their survival skills, they very nearly didn't make it through the Age of Humans. Hunted for their meat and skin, or just shot for sport, saltwater crocodiles had a very close call with extinction. By the 1970s, humans had pushed them to the brink of existence, with only 3,000 of these ancient reptiles left in Australia's Northern Territory. Thanks to action from the Australian Government, they became protected, and they are thriving once more. Though sadly the same cannot be said for other crocodile species.

Platypus

Ornithorhynchus anatinus

Boondaburra • Mallangong • Tambreet • Tohunbuck

Nocturnal

Location	Average Lifespan	Average Size	Average Weight	IUCN Red List Status
Australia	12 years	50cm long	1.5kg	Near threatened

Sometimes referred to as the duck-billed platypus, this wonderfully fascinating and reclusive animal was once thought to be a practical joke when it was first seen by European eyes in 1799, who couldn't believe what they were seeing. The tail of a beaver, pelt of an otter, webbed yet clawed feet, and the beak of a duck. It was known to

Western science as the 'duck mole' for 90 years or so, while men in top hats and stiff collars argued about its existence.

Had they asked the Indigenous population, they would have found their answers much quicker.
The platypus is an egg-laying mammal, known as a 'monotreme'. One of only five left in existence, along with the echidna. The females lay just two eggs at a time, in a burrow dug with her tail. She will then nurse her platypups until they're able to swim on their own. Their handy tails are also used to store fat as a reserve when food is scarce.

Their broad bill feels like suede but is flexible, like rubber, and is used to shovel around the riverbed for insect larvae, small fish and worms. It is a highly sensitive piece of equipment, with built-in electroreceptors that receive signals from movements of prey. They are nocturnal, spending up to 12 hours hunting in the water at night, perfectly designed for their semiaquatic lifestyle, their webbed claws and short, stout legs propel them through water with ease.

They are undeniably adorable, but they hold a secret weapon, venomous spurs. One of very few mammalian species to have venom in its arsenal, the male platypus contains this in a spur on their back feet. Although not lethal to humans, it will cause some pain! Stranger still, the platypus has been discovered to have the ability of biofluorescence under UV lights, another mystery for scientists to uncover.

Great White Shark

Carcharodon carcharias

Witdoodshai • White Pointer • Jaws • Niuhi • Ndiagadar • Hvidhaj

Nocturnal

Location	Average Lifespan	Average Size	Average Weight	IUCN Red List Status
South Africa, Australia, New Zealand, North Atlantic and Northeastern Pacific	40–70 years	4.6m long	1,100kg	Vulnerable

Arguably the most famous shark species in the world after some, well, unfavourable publicity thanks to the 1975 cinema hit *Jaws*, which left quite the bite mark in the minds of the general public. After seeing a ferocious killer shark hungry for humans on the silver screen, many vowed never to enter the water again, at least not until they got themselves a bigger boat.

Since then, conservationists have dedicated their lives to separating fact from fiction to tell the little-known truth about these mighty apex predators, but so much about their lives is still shrouded in secrecy. We know they are very old, existing for at least 70 million years. The average age reported for a great white is between 40–70 years, but we don't know for certain due to the high levels of human hunting and they could, actually, live to be much older if we left them alone.

The largest predatory shark, streamlined for life under the waves, their grey bodies and white bellies are akin to a powerful torpedo that funnels them through the water at 35mph, covering huge distances, travelling from South Africa to Australia and beyond.

One great white, affectionately known as Lydia, gave researchers a glimpse into her watery world. She was fitted with a tracker in 2013, giving away some of her deepest secrets, as data showed her diving to over a kilometre beneath the ocean surface, right to the bottom of the Mesopelagic. This is an underwater realm which exists in constant twilight, home to 95 per cent of all known fish species. Most ocean life hangs out here until night falls on the ocean surface, triggering nightly mass migrations to upper levels of the sea.

This is why most humans spot great whites at dusk or during the night. Great whites will not directly target humans, but we must remember we are a guest in their home when we enter the water and treat it with respect.

Dingo

Canis lupus dingo

Warrigal • Boolomo • Dwer-da • Maliki

Nocturnal and Diurnal

Location	Average Lifespan	Average Size	Average Weight	IUCN Red List Status
Australia	5–7 years	60cm tall	20kg	Vulnerable

Iconic and inquisitive, these wild canines have long roamed the grasslands and foraged in the forests of Australia. At dusk you may be lucky to catch sight of a dingo pack, as they slink into the cover of the night. They are a truly wild species, with fox-like faces and bottlebrush tails. They have a wide-ranging colouration in their fur, with some as red as the Uluru rock, others the colour of the white sands of Lucky Bay. They are an important part of Australia's natural heritage and while technically introduced around 10,000 years ago, they play an important role in the ecosystem, often being a tool to defend nature from invasive species humans so unwisely introduced to the country.

They are not descendants of the grey wolf, like our faithful domestic dogs that we keep as pets, but an East Asian species. They became an important part of everyday life for First Nation people, a kinship that still endures today. They were regular companions as hunting dogs, and were guardians of the camp in both a physical and spiritual sense.

They are more impressive than our domestic breeds, with double-jointed limbs and the ability to rotate their necks 180 degrees, an almost owl-like feat impossible for our pet dogs. They are superb at climbing and can easily amble over an obstacle of 2 metre. They are the Houdini of the canine world.

However, their existence is contentious, with human/ dingo conflicts leading to the construction of the world's longest fence. The dingo barrier fence stretches 5,600 kilometres long and has perhaps become more contentious than the dingo themselves. This was to exclude them from hunting livestock and has had other impacts since.

Sugar Glider

Petaurus breviceps

Nocturnal and Diurnal

Location	Average Lifespan	Average Size	Average Weight	IUCN Red List Status
Australia, Indonesia and New Guinea	9 years	25–30cm long	165g	Least concern

These adorable arboreals are akin to a small squirrel crossed with a miniscule badger. These petite possums have silvery fur, with a black stripe running from their delicate pink and whiskered noses down the length of their spines. They are the epitome of bright-eyed and bushy-tailed, with two enormous eyes filling most of their face, allowing this nocturnal aficionado to see incredibly well at night. The miniscule marsupials can fit easily in the palm of a human hand, but they're concealing a hidden, superhero ability.

Tucked away between their fifth finger and ankle is their patagium, which is a bit like a superhero cape – in reality, it's a fold of membrane that enables this enigmatic animal the gift of nearly flight. Leaping from trees with legs outstretched, this membrane allows them to glide up to 50 metres through the air. That's half the length of the Statue of Liberty if you were to lay her down in New York Harbour. That's a mighty distance indeed when you're a tiny sugar glider.

While gliding, they use their long tails as a rudder to steer them into the paths of insects that they catch mid-glide. During the day, they nest in groups of ten or more, within leafy beds laid down in tree hollows. But it is at night-time that this little forest star shines, built for a life in the natural darkness that its habitat offers.

Tasmanian Devil

Sarcophilus harrisii

Tassie • Purinina • Tardiba

Nocturnal and Diurnal

Location	Average Lifespan	Average Size	Average Weight	IUCN Red List Status
Tasmania	5 years	80cm long	12kg	Endangered

Devilish by name, but are these mighty marsupials really devilish by nature? Short and stocky, they can be seen ambling around shrublands on a scavenger hunt. Their black fur blends them seamlessly into their nocturnal environment and they use an impressive set of whiskers to roam impressive distances of up to 16 kilometres in a single evening. They have excellent hearing, with ears that stand proudly on the top of their big, broad heads, and sensitive noses that sniff out food that ranges from carrion to waste discarded by humans.

They have a monumentally wide mouth, which they are more than happy to show off when threatened, complete with an imposing set of teeth that features two large, sharp canines, which are powerful enough to chomp through carcasses, including the bones. They got their monstrous moniker when Western settlers first heard a blood-curdling howl cry out into the dead of night, their shrieks, snorts and growling in the undergrowth terrified the settler who believed they had found the devil incarnate, and the name stuck.

They are extremely useful as they clean up all the dead stuff in the environment, like an overly ferocious waste disposal worker. They are even known to sleep inside the carcasses they find, so they waste no time when waking in getting right back down to the eating. This also makes it less likely that another Tassie is going to come and steal their meal; as solitary creatures, they really don't like to share.

Seen as a threat to livestock, Tasmanian devils were hunted to near extinction, with farmers on a mission to totally eradicate the animals until they were protected by law in 1941. They are still at risk of being wiped out, but conservation efforts are taking place to keep these fascinating little devils in existence.

Quokka

Setonix brachyurus

Kwoka

Nocturnal

Location	Average Lifespan	Average Size	Average Weight	IUCN Red List Status
Western Australia	10 years	50cm long (excluding tail)	3.5kg	Vulnerable

Australia is home to some of the most charismatic wildlife on the planet, but vying for the crown of most charming is the friendly quokka. These merry little marsupials have become something of a celebrity, known across the globe as the happiest animal on Earth. What makes them so endearing? It helps that they have an adorable appearance, somewhere along the lines of a wallaby crossed with a rat. They have a thick brown coat with two rounded fluffy ears; large, dark and intelligent eyes that were made not only for excellent night-time foraging, but apparently for pulling on the heart strings of humans. They have long back feet and short forearms, just like their kangaroo cousins, and a long, slender tail that has very little fur. But it is their friendly nature that has stolen the hearts of every person to cross their path. They are very trusting, and if given respectable space by humans, they have been known to approach to say hello. It is important not to touch or feed quokka, as this can impact their species negatively.

They are intelligent and capable of following commands and learning things, such as gestures, relatively quickly. These happy herbivores come out under starry night skies to browse the aisles of streamside plants for leaves and grasses, preferring to move on all fours but they can and do bound around and can even climb trees if they need to. They love to socialise and live in family groups, with whom they spend the days snoozing under the shade of trees or prickly plants for protection.

Until recently, there wasn't a collective noun for a group of quokka. The Western Australia tourism board saw their opportunity and launched a competition to remedy this. The winning entry came from none other than surfing legend, Kelly Slater, who offered the term 'shaka', a word synonymous with surf culture. So, while you won't see a quokka in the line-up, you can now refer to a shaka of quokka.

House Mouse

Mus musculus

Nocturnal

Location	Average Lifespan	Average Size	Average Weight	IUCN Red List Status
Global (excluding Antartica)	1 year	20cm (including tail)	20g	Least concern

Scrabbling and scurrying under the protective shield of darkness, this tiny rodent has become one of the most successful mammal species on the planet. Timid as a, well, mouse, this species has become reviled by humans, who ironically have played a crucial role in their success. We fear them, as harbourers of disease and infection, but in reality it is true that where you find humans, you will find mice.

They have coexisted with us from before the dawn of agriculture, using human camps for shelter and food, and it is humans that took them with them, to almost every corner of the globe. They are incredibly intelligent, solving problems with ease, especially where food is involved. Again, living alongside us has made them more intelligent, they have had to get more creative, thinking smarter to avoid us while using us for free food and warmth.

Not only have they been a constant companion for humankind, many of us now owe our lives to this mighty mouse. Its intelligence has meant it is the species most commonly used for biomedical research. Without this intrepid species we would not have cures for polio, meningitis and many cancer-fighting drugs. All of a sudden, those morsels of food don't seem too high a price to pay for medical improvements.

They are a communicative species, using squeaks and scents to share information. Their excellent hearing can detect frequencies undetectable to the human ear. But one thing that may surprise you is that they sing. The melodic male house mouse has been found to sing songs individual to them to attract mates; females also have their own repertoire. Their songs cannot be heard by human ears, reaching the extreme falsetto heights of the ultra-sonic range. Maybe Cinderella wasn't so far-fetched after all.

Kākāpō

Strigops habroptila

Owl Parrot • Moss Chicken • Night Parrot

Nocturnal

Location	Average Lifespan	Average Size	Average Weight	IUCN Red List Status
New Zealand	60 years	60cm long	2kg	Critically endangered

Waddling slowly along the ferns and fungi of the New Zealand forest floor, a genteel and unhurried creature browses for seeds and plants among the fronds with the air of an eccentric Victorian gent, promenading with hands clasped behind its back, perusing the local botany. You could not be blamed for mistaking it for a moving mossy rock, until its head turns and you are faced by two inquisitive, coal-black eyes either side of a large grey beak. Let me introduce you to this truly noble soul: the kākāpō.

His credentials are many, including two distinct titles as both the heaviest and the world's only flightless parrot. He also has a theatrical flair with a touch of the operatic. Kākāpō are the only parrot species to engage in a 'lek', which finds males banding together to prepare an auditorium by clearing an area of plants before each scraping their own individual stages, known as a bowl. Scene set, the show begins. Each male will compete to be the most accomplished bass, inflating their throat sacs before releasing a booming call that resonates across the romantic setting of the dark and starry night to attract flocks of, hopefully, swooning female fans. Once upon a time, they would have been part of the chorus of a summer's night, but our poor, dear kākāpō is almost facing its final curtains.

Just 247 of these enigmatic creatures are left in existence. Before humans arrived on the islands, it was a widespread bird and had no land-based predators; it sacrificed its power of flight in favour of strong legs. Brilliantly suited to climbing trees in search of the ruby-red fruits of the rimu tree, the kākāpō can, at a push, glide but, realistically, this is more akin to it just falling with style.

Hunted by Māori for its meat and feathers, it is considered a *toanga* (a sacred treasure). They emit a sweet scent, like honey, making it easily found by introduced predators, such as dogs. Efforts to save this iconic species are in full swing, so that the show can truly go on.

European Hare

Lepus europaeus

Brown Hare • Race-the-wind • Dew-flirt • Ysgyfarnog • Mawkins • Maigheach

Nocturnal

Location	Average Lifespan	Average Size	Average Weight	IUCN Red List Status
Europe, North and Western Asia (Native)	4 years	58cm long (excluding ears)	4kg	Least concern

On a moonlit March night, as a rolling mist drapes itself within dips and across ditches, flickering shadows chase themselves through the shimmering grasses. A kindle of hares weave between the whirling winds, slipping in and out of sight, ungoverned by normal natural laws. Trapped in the Moon's gaze, tawny eyes observe their night-time landscape – illuminated, surveying, all-knowing. Long elegant ears are tipped with sooty black and create a striking silhouette. Look too closely and she evaporates like the morning dew.

The hare has long been a symbol of the mysteries associated with the nocturnal. The unpredictable nature of the hare has seen it linked to the Moon throughout the ages, and just like the Moon's tendency to appear unexpectedly in the sky, the hare too can surprise onlookers. The hare has attended many gods in many cultures, including Freya, the Norse goddess, whom travelled with a sacred hare; Kaltes, a shape-shifting Moon goddess from Siberian folklore, often appeared in the form of a hare; and Cerridwen, in Cymru, has links too.

It has been introduced to countries outside its natural range, such as Australia, where it is sometimes considered an invasive species. In the United Kingdom, it is the fastest land mammal, reaching speeds of 40mph. Spring is a good time of year to watch for hares, and if you're lucky, you may see them engage in 'boxing', including high-speed chases and the exchanging of blows, which is where we get the saying 'as mad as a March Hare' from. It may look like chaotic madness, but it is in fact part of the courting rituals of this otherworldly creature.

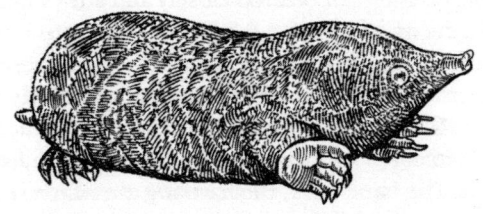

European Mole

Talpa europaea

Twrch Daear • Umpty-tumpt • Mouldiwarp

Nocturnal and Diurnal

Location	Average Lifespan	Average Size	Average Weight	IUCN Red List Status
Europe and Western Asia	3 years	16cm long	110g	Least concern

A velveteen coat of the finest black fur comes to its end
at a sensitive pink snout not dissimilar to a pig's in
miniature form. With no visible ears and two minute
eyes, this mining mammal is built for life under the
earth, in complete darkness. Its most notable feature is
its impressive set of shovels, which we call feet. They are

disproportionately massive in comparison to the rest of its body and equipped with five sharp claws, giving this subterranean species some serious engineering capabilities. They can tunnel at a rate of 20 metres per day, shifting an impressive 6 kilograms of earth in just 20 minutes.

Sadly, most of us will never meet a mole as they prefer to stick to the depths of the soil. The closest we are likely to come is if we were to interact with a fresh mole hill. The bane of anyone seeking a uniformly neat lawn, the soil they throw up when digging through the earth is actually an excellent source of sterile potting compost. They live in a tunnel system under the ground, and despite what you hear from many gardeners, they are excellent for soil health, aerating it and improving drainage.

They also do a large amount of pest control. As insectivores they're always on the hunt for their next tasty meal, using the sensitive bristly hairs on their body to feel for the vibrations of their would-be prey. Living underground in complete darkness they have no need for good eyesight, instead they rely on sensitive hairs to navigate the tunnels and seek out insects living under ground. They particularly enjoy a good earthworm, eating around twenty each day, which equates to just over half its body weight. It captures insects in its tunnel system, using it to trap things like leatherjackets. Moles are venomous and use toxins found in the saliva to paralyse its prey and keep it fresh for a feast later on. But don't worry, they can't paralyse humans.

Caracal

Caracal caracal

Desert Lynx • Gazelle Cats • Persian Lynx

Nocturnal

Location	Average Lifespan	Average Size	Average Weight	IUCN Red List Status
Africa, India and Asia	12 years	42cm tall	18kg	Least concern

A powerful predator pads across the Eurasian Steppe, picking their way through the vast grassland in search of prey to stalk. The long legs of this commanding creature carry it swiftly throughout the endless dark of the night. This solitary beast can wander over 20 kilometres each night in the search of sustenance. Perhaps a small antelope, a solitary hare, or maybe they will propel themselves into the branches of a nearby tree, to claw at anything that mistook the trees canopy as a safe place to sleep.

The caracal is an impressive, if not slightly odd, looking beast. The body is that of a lioness – powerful, muscular, built for speed and strength. Its head is more akin to the size and shape of a domestic cat, until of course you get to those fantastically flamboyant ears. Their coat is a short and dense combination of cream-coloured fur that stretches from around the stomach before blending seamlessly into a rustic tan. Their faces are decorated with black lines and white markings that highlight the imposing stare of this refined, regal feline. Its ears are crowned with two gloriously long tufts of black fur, feathered upwards over 5cm from the very pointed tip of the ear. This fur isn't just a fashion statement, as it comes in very handy for camouflage, breaking up the sharp shape of this cat's outline among the soft, withering grasses. It is also thought that these tufts can be used to communicate with other caracals, twitching transmitted messages across the rolling plains.

'Caracal' comes from *karakulak,* which means 'black ear' in Turkish. In India and Persia (modern-day Iran), these cats were semi-domesticated and used for hunting birds. They are capable of leaping great heights, and competitions were held to see how many pigeons one cat could bring down mid-flight. It isn't difficult for caracals to swipe ten or more birds at once, bringing about the old saying 'to put a cat among the pigeons'.

Common Wombat

Vombatus ursinus

Bare-nosed wombat • Wambad • Tasmanian Wombat

Nocturnal

Location	Average Lifespan	Average Size	Average Weight	IUCN Red List Status
South-eastern Australia	15 years	98cm long	26kg	Least concern

Chunky, chubby and cuddly, the wombat is a magnificent marsupial that can scoot around the Australian scrub at a surprising speed given their hefty and stout stature and are, essentially, a cute and furry lawn mower. They spend their nights grazing away at anything in their path –

grasses, shoots, leaves. As the Sun rises and their feasting finishes, wombats head to the relaxing cool shade of their underground burrows, where they doze away the day, often with all four legs up in the air. The common wombat is one of three wombat species, alongside the northen and southern hairy-nosed wombats.

When first encountered by Western settlers, they believed them to be a type of badger. Resembling a small bear, their closet relative is actually the koala, but wombats don't live in the treetops. Ambling around like four-legged weebles, they mean business when it comes to digging: five toes are kitted out with long, strong claws that can shift a serious amount of soil. They live in underground tunnel networks known as warrens that can be over 200 metres long, usually housing around eight wombats in a group known as a wisdom. They are very wise indeed, making their homes underground offers them protection from both predators and the heat, including bush fires, and they are not the only animal to benefit. During wildfire events, other small animals, from wallabies to birds, will flee to the safety of the wombat warrens until the flames subside, making these wise little souls the heroes of the bush.

The name wombat is taken from Darug, the indigenous language spoken by the Traditional Owners of Sydney. If you want an excuse to celebrate the joys of wombats, mark October 22 in your diaries, as World Wombat Day!

Starry Night Octopus

Callistoctopus luteus

Night Octopus • Smallspot Octopus

Nocturnal

Location	Average Lifespan	Average Size	Average Weight	IUCN Red List Status
Western Pacific – Indonesia, Japan, Taiwan, Hawaii	3 years	70cm long (including tentacles)	12g	Least concern

A living treasure chest glides gracefully along the sands of the Pacific seabed. Gilded in pearls and diamonds, their skin shimmers like a bejewelled night sky on the darkest, clearest summer evening, twinkling in the glimpses of light that reach through the reef. A shoal of small fish glimmer in the twilight and tropical waters, unaware that a watchful eye follows their every move. Making a move from the darkness with skin glowing like the finest orange silks that slip between the coral branches, lighter than air and slicker than water, a long arm unfurls before revealing an array of small saucers, ready for supper. The school of fish loses a student. Before they have chance to scatter, eight arms wrap around the solitary soul that suddenly vanishes from sight, sequestered in a castle of coral.

It is clear to see how this octopus got its name. A beautiful creature with bright red or orange skin, they are unlike many other octopus species in that they do not completely change colour, only flashing its 'stars' as white spots on the skin's surface when it feels the need to communicate to others in its surroundings, primarily when it feels threatened. The white spots need to be seen to be believed and positively glow on the surface of the cephalopod. With its eight tentacles swirled around its mantle, you could truly believe it was a galaxy like our very own Milky Way come to life on the ocean floor.

This species can be found in coral reefs or rocky areas where it can easily hide away from prying eyes. It is particularly solitary and prefers to emerge only at night to avoid any competition for food while it hunts among volcanic rocks.

Tawny Frogmouth

Podargus strigoides

Aleddjumud • Djurrul • Kuluyhkuluy

Nocturnal

Location	Average Lifespan	Average Size	Average Weight	IUCN Red List Status
Australia	14 years	50cm long	680g	Least concern

Many strange sounds carry across the Australian night, and adding to the nocturnal orchestra is the steady pulsating call of this peculiar perching creature. Through the darkness you may just catch the glimmer of large eyes the colour of the yellow wattle plant. By the silver light of the Moon, you may just glimpse the mottled grey feathery outline of what appears to be a short owl with a large, squished head. This feathery fascination is no owl, but the phantasmal tawny frogmouth.

By day, you stand only the slightest of chances in spotting this chameleon of the canopy. That being said, you will see it without realising what you're seeing. So cryptic is its colouring that it blends in near-perfectly with the branches of the stringybark tree. They roost during the day, stretching their necks to the sky and closing their eyes, becoming one with their surroundings. They hunt at night, which is when you will see that they are aptly named thanks to their cavernous mouths, which open wide to collect as many insects as possible.

Fiercely territorial, once they have moved into a home they like, they will remain there, sometimes for more than a decade. Loyal, too, these birds mate for life. A wildlife researcher witnessed the sorrow of a widowed female, observing her as she made a morose weeping call. She refused to take a new mate for a number of years after her partner's death, leading experts to believe these birds may grieve for their lost mate.

Sadly, they are susceptible to death on the roads as they are insectivores and attracted to the flying insects that flock to car headlights. This rarely ends well for the bird. They have adapted to use streetlights at night to forage for the insects that gather in their cones of light. Great for the tawny frogmouth, but not so much for the insects where the draw of these lights is affecting the precariously weighted ecosystem.

Mexican Free-tailed Bat

Tadarida brasiliensis

Brazillian Freetailed Bat • The Guano Bat • The Mastiff Bat

Nocturnal

Location	Average Lifespan	Average Size	Average Weight	IUCN Red List Status
North, Central and South America	8-11 years	10cm long with a wingspan of 33cm	12g	Least concern

On a balmy evening in Austin, Texas, the Sun is beginning to set over Lady Bird Lake. You find yourself drawn towards a crowd of hundreds of people, all eagerly lining up along the Congress Avenue Bridge. Your interest is piqued as you sense the excitement as more people are drawn up to the deck. To your surprise, the river has also become crowded with aquatic admirers of the bridge. Paddleboarders, kayaks, and boats full of people standing shoulder to shoulder, all craning to get a good look at the bridge's arches that reflect the colours of sunset in the water. People start to gasp and point, 'Bats!', you face east and your eyes are met with the most unforgettable sight. A steady stream of black, busy shadows have taken to the air, mingling with the delight and joy of their audience below. The stream becomes a torrent, a crepuscular crescendo as 1.5 million bats flood from their unlikely home into the skies above the city lights.

This bridge is home to the largest urban colony of bats on the planet. Engineering works in the 1980s unintentionally upgraded the bridge to make it the perfect home for the Mexican free-tailed bat. One of the most abundant mammals in North America, their nightly display in Austin has seen them become true stars of the night, with over 100,000 people making their own special migration to witness the spectacle.

Bat species account for over 20 per cent of all the planet's mammals but this particular bat, with its huge ears and long tail, has a particularly impressive skillset. Migrating from Mexico to North America and back again each year, their migrations are so monumental that they can be detected on weather maps. They may be small, but that doesn't stop them flying up to 10,000 feet high in search of food.

Greater Bilby

Macrotis lagotis

Mankurr • Ninu • Rabbit-eared Bandicoot • Nyarlku
• Dalgyt • Pinkie

Nocturnal

Location	Average Lifespan	Average Size	Average Weight	IUCN Red List Status
Australia	7 years	20–55cm long	2.5kg	Vulnerable

Half-hopping, half-crawling, and sometimes galloping, the greater bilby is a marsupial icon of the Australian night. Imagine a wallaby crossed with a rabbit and you may come close to envisioning the character that has taken up the mantle of the Easter Bunny for many children across Australia. Giant, rabbit-like ears can make up as much as 66 per cent of its body length; its silky fur, which would make a chinchilla jealous, peters out into a long, pink, hairless snout that is all the better to sniff out insects and seeds, which it searches for under the starry night skies. It uses its front paws like hands to gather food. Do not be fooled by its small size, they can quite literally gallop, holding out their long tail behind them, taut like a black banner with a flick of white at the tip like a small crescent moon. Their huge ears make up for their poor eyesight as they are very sensitive to light.

They are valued as ecosystem engineers, with a single bilby capable of moving an impressive 3 tons of earth a year during the construction of its elaborate underground home – spiral-shaped burrows that lead deep into the protective ground, which shields them from both predators and the heat of the day. They are also quite partial to sleeping on their head and, like the wombat, are quite willing to share their homes with other wildlife during forest fires.

The greater bilby was once found right across the wilds of Australia. They are thought to have existed for some 15 million years and, until Europeans settled on the continent, could be found roaming around 70 per cent of the mainland. That's now been reduced to a mere 20 per cent and is continuing to decline with concerns rising that this much-loved character of the Australian nightscape may be lost forever, with extinction looming on the horizon.

There are projects underway to help save the greater bilby before it goes the same way as the lesser bilby, which sadly became extinct in the 1950s. Indigenous rangers are now a crucial part of the plans to save this enigmatic darkness-dweller.

Chinchilla

Chincilla lanigera

Chilean Chinchilla • Coastal Chinchilla Long-tailed Chinchilla

Nocturnal

Location	Average Lifespan	Average Size	Average Weight	IUCN Red List Status
Western South America	10 years	20–55cm	600g	Critically endangered

Chinchillas are perhaps most well known for their fur –
luxurious and velvety soft, the thick grey coat is perfect
for keeping them toasty when foraging during night-
time feasts upon the chilly slopes of the Andes. Their
fur has seen them targeted by humans for their valuable
pelts, used to adorn the backs of the rich and famous

around the world. This trade for fashionable furs has led to this community-minded critter becoming close to extinction in the wild.

Left to their own devices, the chinchilla is very sociable and lives in colonies that can number into the hundreds. They are most active at night, when they come out to forage under the cold and clear skies of the Chilean mountains, where temperatures can plummet to lows of -5°C . A type of large rodent, they are extremely agile, capable of leaping up to 1.8 metres in the air. As a prey species, they look after one another when out in their feeding grounds. One will always be a lookout. If a predator approaches, the cute, cuddly chinchilla can soon put on a frightening appearance, standing on its hind legs and making their long bushy tails even bushier to appear as intimidating as possible.

They like to sleep the days away in burrows or rocky crevices, but they will venture out for a bit of luxurious treatment, sunbathing and partaking in dust baths, during which they roll around and cover themselves in the volcanic ash present on the hillsides to keep that coat in glorious condition. They are one of the longest living rodents and can reach 20 years of age in captivity.

Eastern Barred Bandicoot

Perameles gunnii

Warron • Peek Woorroong •Wydung •Wateun

Nocturnal

Location	Average Lifespan	Average Size	Average Weight	IUCN Red List Status
Tasmania and mainland Victoria	3 years	35cm long (excluding tail)	700g	Vulnerable

Living out solitary and quiet lives, these meek and mild marsupials look akin to a rabbit but with a long, rodent-like snout, complete with whiskers and a delicate pink nose. They have stubby, thin tails and strong claws that are perfect for digging. As their name suggests, they have prominent markings made up of buff- or cream-coloured stripes that stand out against its soft, brown coat.

Like many burrowing species, they are extremely valuable as ecosystem engineers, keeping soils aerated and healthy. They nest in shallow, conical pits that turn over fresh soils. This may not sound overly important but by turning over the soils they are digging in dry materials like dead leaves, and grasses are removed from the surface, which helps reduce the number of flammable materials on the ground. A very useful tool when you live somewhere that is increasingly susceptible to wildfires. Over a year, the digging of one bandicoot moves an equivalent of an adult elephant's weight in soil.

They are wholly nocturnal, leaving the safety of their homes (digs) to forage among the grasslands and forests for tasty insects and fallen fruits. By keeping to the hours of darkness, they are better protected from the prying eyes of predators and the hot temperatures of the day.

If you are of a certain age, you may have already encountered this creature whether you realise it or not. For those of us partial to a bit of PlayStation in the late 1990s and early 2000s, you probably played *Crash Bandicoot*. The main character is actually an eastern barred bandicoot, who was snatched from the wild and experimented on, which is why he neither looks like nor has any of the characteristics associated with the bandicoot.

Being immortalised as a game character hasn't helped these sweet and shy beings avoid dwindling numbers but conservation efforts are underway, which has already seen their Red List status shift from endangered to vulnerable.

Night Parrot

Pezoporus occidentalis

Porcupine Parrot • Midnight Cuckoo

Nocturnal

Location	Average Lifespan	Average Size	Average Weight	IUCN Red List Status
Australia	10 years	23cm long	104g	Critically endangered

A ghost of Australia's past has come back to haunt the world of conservation. A natural treasure goes about its nightly adventures, unseen, unheard and unknown. A small army of people have been searching high and low, over thousands of miles to prove its existence. This mysterious bird managed to elude humans for over a century and was presumed extinct until very recently. A gem prized by Victorian collectors, this emerald-green parrot with ebony spots has been shrouded in mystery and controversy since it vanished over a hundred years ago.

Very little is known about how this bird lives its life, and we don't even know how many currently exist in the wild today with estimates ranging from 40–200 individuals. Huge efforts have gone into trying to safeguard this precious parrot from a second extinction, but scandals abound: fake recordings and photographs have left the conservation world shocked, but it hasn't deterred the prospecting ecologists who are determined that this bird must once again thrive.

A rare opportunity to save an extinct species doesn't often come around, but with a country as huge as Australia, it has been more like panning for gold in an ocean rather than a stream. Glimmers of hope have come in the form of sparse sightings, as this elusive parrot appears to inhabit some of the harshest climates on the planet, where the heat is intense and water scarce. It has been found clinging on to existence in clumps of old spinifex, a spikey shrub, likely leading to its colloquial name of 'porcupine parrot'.

As the name suggests, it is a nocturnal bird. However, it appears to have some difficulty navigating at night as its eyes are not equipped with good night vision and it seems to have an unfortunate habit of flying into things in the dark. Hopefully, conservationists will one day be able to reveal to the world that this precious nocturnal treasure is a rarity no more.

Amami Black Rabbit

Pentalagus furnessi

Ryukyu Rabbit • Amami no-kuro-usagi

Nocturnal

Location	Average Lifespan	Average Size	Average Weight	IUCN Red List Status
Japanese islands of Amami Oshima and Tokunoshima	10 years	41–51cm long	1.3–2.7kg	Endangered

Around 800 miles from the hustle and bustle of Tokyo, the world's largest city, lies a small Japanese island. It may as well be an entirely different place in time to the futuristic capital. Deep in the heart of its beautiful, lush green forests that shroud the mountainous hillsides, hides an elusive and shy species, a relic of evolution, that has come to be known as 'the living fossil'.

Stout and sturdy, they look more like a wombat than a rabbit. Its dark fur lost completely to the shadows of the forest. It has a broad nose; short, rounded ears; and a whisper of a tail. When we think of rabbits, we see them frolicking through open country, fields and meadows, but this rarity lives in dense and ancient forests, making its home among mossy boulders and fern-framed burrows. They have survived dangerous predators, including venomous snakes, which may have led to the curious behaviour in which when a new litter is born, the mother will build a new nest – away from her burrow – in which to keep her newly born kits. After feeding, she seals the burrow up, heads out to feed herself, and only returns once every 2 days to avoid predators locating her young. Given their clandestine, nocturnal lifestyle and existence in surreptitious surroundings, there is still much to learn from these unique rabbits.

A living chapter from the story of evolution, the Amami black rabbit retains some characteristics found in the ancient species of rabbit that existed over one million years ago, a long time before Homo sapiens, which evolved as a species and has made life very difficult indeed for this creature. By studying species like these, we can learn so much about how life on our planet has evolved. They are of such importance to the Japanese government that they declared them a Natural Monument in 1921, giving them protected status.

Giant Panda

Ailuropoda melanoleuca

Mò • Flowery Bear • Zhúxióng • Bamboo Bear • Big Bear Cat

Nocturnal and Crepuscular

Location	Average Lifespan	Average Size	Average Weight	IUCN Red List Status
Sichuan, Shanxi and Gansu Provinces, China	20 years	1.8m long	113kg	Vulnerable

If any animal needed no introduction, it would most likely be the giant panda. They are China's *guabao* (national heirloom) – a symbol of peace and friendship. They are also the face of the world's largest conservation charity, the World Wildlife Fund (WWF). They have even been used as furry delegates, representing China around the globe, forging international friendships and stealing hearts across the world. They have become synonymous as a symbol for nature conservation and a message to mankind that we must live in harmony alongside the incredible wildlife we share this planet with.

The giant panda diet is around 99 per cent bamboo, which is not an overly nutritious meal and would require a fully grown adult to eat around 35 kilograms per day to sustain itself. Human interference, however, has meant there is no longer enough bamboo left in the forests to sustain these animals as their habitats shrink. All is not lost as there are now over fifty panda reserves protected by the Chinese government that aim to give their star species' numbers a boost.

In the wild, giant panda are unusual as they are not truly nocturnal, diurnal or crepuscular like most other species. They seem to have their own rhythm, with activity peaks during the day, middle of the night, and just before dawn. They are built for nocturnal foraging; their eyes have cat-like slits instead of the rounded pupils typical of other bears, which gives them a highly developed night vision. Their eyes have also given rise to one of their many nicknames, 'Big Bear Cat'. Unlike other bear species, they don't hibernate during the winter, despite living in very cold mountainous regions up to 11,000 feet above sea level.

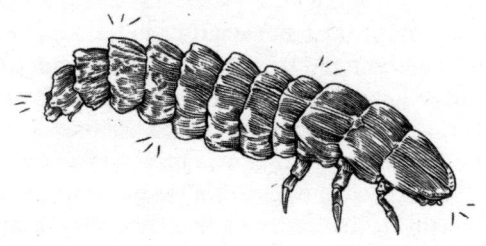

Glow Worm

Lampyris noctiluca

Light of the Hedge • Magien • Tân Bach Diniwed
• Cuileag-shnìomhai

Nocturnal

Location	Average Lifespan	Average Size	Average Weight	IUCN Red List Status
Europe and Asia	2 weeks (adult stage)	25mm	104g	Near threatened

Enjoying a walk on a balmy summer evening close to the summer solstice, the light is only just starting to fade away and bats start to feast on the moths that flit around the corridors of hedgerows. Above you a thin sliver of the new crescent Moon perches delicately upon the purple banding of the horizon. The hedgerows are shrouded in darkness, the road is an empty abyss where light no longer reaches the ground. A large moth rustles the leaves of a honeysuckle and your eyes are drawn to the shadowy grasses that fringe the lane's edge. You pick out glowing lights, strung about the strands of grasses like glimmering baubles of yellow and green. You would be forgiven for briefly believing that fairies had made their home in the hedgerow but, leaning closer, you discover this to be a series of little glowing bugs: meet the glow worm.

Their name is misleading – for starters these are beetles, not worms. After having spent nearly 2 years in a larval state, feasting on snails and slugs, the female is now transformed into her final and most magical form. When conditions are just right, she will climb to the tops of tall grasses and set to glowing. From her lofty position she can be more visible to the males she is hoping to attract. Male glow worms fly, and they locate females by looking for the tell-tale glow, like a shining beacon guiding him safely home. The females can glow every night for around 10 days, and will only stop glowing when they have found a mate. Once she's successfully laid her eggs, her lifecycle is now complete.

Sadly, glow worms are in decline across the UK with recorded numbers down three-quarters since 2001. The males can be attracted to artificial light sources, meaning they could fail to mate, which affects the numbers born into the next generation.

The glow worm belongs to a family called Lampyridae, which translates from Greek to 'shining ones' and 'noctiluca', latin for 'something which shines by night', so loosely their scientific name translates to 'night lamp'.

Manx Shearwater

Puffinus puffinus

Aderyn Drycin Manaw • Devil Bird • Âmes damnées
• Cánóg dhubh • Púicín Gaoithe • Wind Puca • Scraayl

Nocturnal and Diurnal

Location	Average Lifespan	Average Size	Average Weight	UK Conservation Status
North Atlantic islands and South Atlantic (migratory)	15 years	30–38cm long	430g	Amber

Your wooden boat finally comes ashore in the fading light of the day. Grateful to have land under your feet, you set up camp on this strange isle and look forward to your first night tethered to terra firma alongside an exhausted crew.

Sleep comes quickly, even quicker than darkness until you are abruptly wrenched from your slumber by thousands of cackling voices that fill the air, a raucous racket that sends shivers down the spines of each and every sailor. 'Devil bird', a terrified voice whispers as you rush to push the boat back out to sea and fling yourself into the safety of the hull, welcoming the waves that push you away from the haunting calls of tortured souls.

For hundreds of years the Manx shearwater was feared by sailors and those living on remote islands. Their throttled cries sound breathless, like a rooster struggling for air, and has led to myths and stories that these birds were the tortured souls of the damned, drowned sailors, pirates or even the devil himself. We are a bit dramatic as a species, aren't we?

Seafaring specialists, they're incredible flyers, 'shearing' along the faces of waves with their sooty black wings stretched taut, rolling with the waters, giving flashes of their white bellies to the sky. They congregate in rafts of thousands of birds, anchoring themselves by remote islands, waiting for their time to come ashore. While the sea is their domain, they must return to solid land to breed. So vulnerable to life on land, they will use only the cover of the darkest nights to disembark en masse, guttural cries filling the darkness as they return to find their burrows. They lay their eggs underground, another safety measure against landlubbers. When time to fledge, the young must find their own way out into the endless night and navigate their way to South America. A Manx shearwater will fly over 5 million miles during its lifetime, which is the equivalent of travelling to the Moon and back ten times.

Colombian Night Monkey

Aotus lemurinus

Owl Monkey • Gray-bellied Monkey • Musmuki

Nocturnal

Location	Average Lifespan	Average Size	Average Weight	IUCN Red List Status
South and Central America	20 years	34cm long	1.3kg	Vulnerable

In the evergreen tropical forests of South America, night has fallen over the jungle, but this doesn't mean everyone is sleeping. There is a bustling nightlife from the forest floor to the tops of the trees, whirring, hooting and screeching away. High up in the colossal canopy of the Colombian Amazon, there's a chattering, and a clattering.

The only truly nocturnal species of monkey, night monkeys leap from tree to tree and use the light of the Moon to hunt for insects or nectar to forage.

To take a closer look, you would have to scale to the heady heights of one of the towering trees to find a night monkey nest in the hollow of a trunk or nestled among vines thicker than rope, as they rarely ever come down to the forest floor. As you climbed, perhaps a pair of curious, giant amber eyes would peek down at this strange intruder. Their bushy white eyebrows frame their small, rounded faces. Three black stripes trace the top of their heads, skirting around their barely visible ears. The wide face and huge eyes immediately call to mind the features of an owl. A grey woolly coat gives them protection from insect bites and the elements. A long tail, which unlike other monkeys is not prehensile and cannot be used for holding things. These beautiful and social creatures are at risk due to being illegally taken from the wild to be used in medical research and other threats.

Colombia is the second-most biodiverse country in the world. It's not just biodiverse, it's megadiverse; home to over 63,000 species amounting to an astonishing 10 per cent of all flora and fauna species on the planet. Over 3,600 of these species exist only in Columbia, for some perspective on that, the UK has just 100 endemic species. However, the survival of all this life hangs in the balance, as deforestation for things like logging, mining, agriculture, and other illegal activities threaten all this life. There are glimmers of hope that this may be turning around, with deforestation figures down on previous years in 2023.

Death's-head Hawkmoth

Acherontia atropos

African Death's-head Hawkmoth • Le Sphinx Tête De Mort
• Doodshoofmot • Conach na cealtrach

Nocturnal

Location	Average Lifespan	Average Size	Average Weight	IUCN Red List Status
Europe, Africa	6 weeks (as adult moth)	12cm	2.8g	Not assesssed

This moth is truly the rebel of the moth world. Such an anarchist, it nearly brought down the British monarchy by merely daring to exist. A migrant visitor to UK shores, it is the largest moth that frequents those lands in the late summer months. Not only that, it is believed to be the fastest moth in the world, reaching speeds of 30mph.

If it were a human it would be dressed in leathers and ride a Harley-Davidson, while listening to death metal (maybe some Meatloaf). Fully nocturnal, like a moth out of hell he'll be gone when the morning comes.

It isn't difficult to see where this creature got its very metal name. When its wings are closed, it is cloaked top to toe in the most sumptuously dark brown and black velveteen 'fur', which is actually made up of an intricate mosaic of tiny scales. But its thorax (the bit behind their head) is emblazoned with yellow markings that depict a human skull. When its wings open, like a roackstar wearing a cape, it reveals a flashy set of yellow hindwings and black-and-yellow-striped abdomen. It's got the vocals to match, with the unnerving ability to emit a falsetto squeal when disturbed.

Like many frontmen, it's developed a taste for the finer things, opting to feast on honey rather than nectar. It's even got its own VIP access to the hives of honey bees. By emitting an intoxicating pheromone, similar to one bees produce, it can get to places other moths could only dream of. Once in the hives, it gorges itself on honey.

Naturally, these behaviours led to humans being rather suspicious about this being. It has been long associated as an omen of doom, or even death itself, coming to whisk your soul away. In Greek mythology, Atropos was responsible for the cutting of the fine thread that would end the life of mortals. In regency England, poorly King George III was tormented into a bout of madness by the little devils visiting his bed chambers at Kew, threatening the stability of his reign. They clearly had it in for the nobility, as they were reportedly sighted at the execution of King Charles I.

Brown Bear

Ursus arctos

Mathan • Medved • Arth • Bjorn • Akiak • Mesikammen • Dubb

Nocturnal and Diurnal

Location	Average Lifespan	Average Size	Average Weight	IUCN Red List Status
North America, Canada, Europe, Asia, Russia	20–25 years	2m long	350kg (male)	Stable

Brown bears have captivated humans throughout our history, from ancient folklore and bear gods to the teddy bears we give to children as a bedtime companion. They can be quite like us in many ways, using their paws like hands to hold objects and eat, and their ability to walk upright on their two hind legs. They cover a wide range of the planet, but their territory is vastly reduced, now populating just 2 per cent of their former range. They are one of our largest carnivores, although truly, they are omnivores, eating a wide and varied diet depending on what is available to them (including human rubbish).

Something about their strength and tenacity has always inspired us. They are the national animal of Finland, where in local folklore a bear known as Otso was said to be the king of the forest and all the animals within it.

Brown bears can be diurnal, but they appear to be primarily nocturnal in Europe, where they exist in closer proximity to humans. Bears will avoid conflict where possible, using the cover of the night to stay away from the danger of humans.

They are, of course, also famed for their ability to hibernate. In Finland they retreat from the scarcity of winter into their cozy dens from October to March, after consuming lots of food to build up fat reserves that they can use over winter, then emerging again when food returns.

Humans have even immortalised the bear in the heavens with not one, but two constellations – Ursa Major and Ursa Minor, the Greater and Lesser Bear, respectively. This story goes back to prehistory, linked to the ancient tales of the Cosmic Hunt, over 13,000 years ago. It has been carried all the way through to the Romans and Greeks who linked the story of Callisto and her son Arcas to the constellations, as Zeus turned them into bears and sent them into the sky to forever wander around the pole star together.

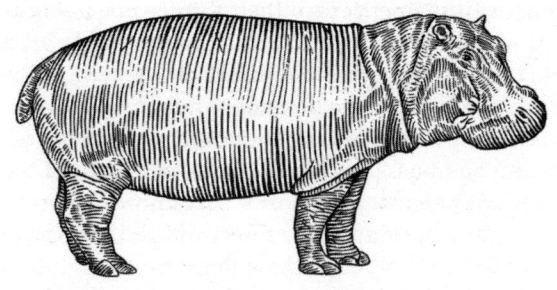

Common Hippopotamus

Hippopotamus amphibius

Kubu • Invubu • Kiboko • Imvubu • Seekoei

Nocturnal and Diurnal

Location	Average Lifespan	Average Size	Average Weight	IUCN Red List Status
Sub-Saharan Africa, Tanzania, Kenya, Zambia, Zimbabwe, Uganda	36–40 years	3.4m long	2,000–3,200kg	Vulnerable

A hulking giant of the African wetlands, one of the planet's only remaining megafauna is the unmistakable common hippopotamus. These semi-aquatic mammals are a mass of boulder grey that bob about in lakes and rivers in large social groups, usually of around forty or so, known as a pod or bloat. Their skin is extremely thick at 6 centimetres, and pretty much hairless. To protect themselves from the African sun, they secrete an oily substance known as 'blood sweat', which turns a red colour, and acts much like a sun cream by shielding their skin from the UV rays.

Their giant, peanut shell-shaped heads are nearly always totally submerged in water, their eyes, ears and nostrils all sit on top of their face so they can breathe easily while wallowing. They have strong but stubby legs, like granite pillars, that end in four webbed toes that help propel them through the water, pushing along the bottom of the riverbed like a chubby submarine. Despite not being great at swimming, they can hold their breath for an impressive 5 minutes between breaths.

Their scientific name translates as 'river horse', but they are no relation to equines. Their closest relatives are actually our cetaceans (whales, dolphins, porpoise). These giants are mostly nocturnal, leaving their watery surroundings to peruse the grasses just after sundown.

Despite being herbivores and eating vegetation, they have a terrifying set of teeth. If you get too close to a hippo, it will happily reveal one of its most impressive skills: throwing its head back and opening its jaws over a metre wide to show off sharp, gargantuan canines that can grow up to 50 centimetres in length. They may be cute to look at, but hippos are very aggressive and will attack if they feel threatened. Despite appearances, they can reach speeds of up to 30km/h on land. With a biting force three times that of a lion, I wouldn't want to surprise a hippo!

Hippos would have been widespread in prehistoric times, with remains of some ancestors even being found under Trafalgar Square. But they are now at risk of extinction due to habitat loss and illegal poaching.

Nocturnal African Dung Beetle

Scarabaeus satyrus

Scarab Beetle • Miskruier • Tumble Bugs • Ibhungane Lezinyosi • Inkuba

Nocturnal

Location	Average Lifespan	Average Size	Average Weight	IUCN Red List Status
South Africa	3 years	2.5cm long	4kg	Not assessed

Scarabaeidae is a huge family of somewhere in the region of 35,000 species of beetles. They come in a wide range of colours – some beautifully metallic, some kaleidoscopic – and quite the range of sizes, from 0.5 centimetres to a Herculean 16 centimetres. As a family, they have travelled across the continents, and can be found everywhere on the planet, bar Antartica. Short and stout, they have squat, wide heads the width of their hard carapace covered bodies, that protects them like a sarcophagus, from under which six long legs protrude, equipped with specialist scarab shovels to dig into dung and soil. They use these to gather up dung into balls, which they roll away, sometimes burying them, to be used as a food store or a breeding chamber.

To the ancient Egyptians the scarab was sacred, symbols depicting the beetles can be found on everything from jewellery, such as amulets, and decorations through to scripture and funerary items. To them, the dung beetle's 'globes' represented the Earth, which the dung beetle kept spinning. They were intertwined with the god Khepri, who was responsible for the renewal of life and the movements of the Sun.

But this particular species has surprised scientists by demonstrating a more direct link to the heavens. In 2013, a study by Professor Marie Dacke showed that during its dung rolling in the darkness, the nocturnal dung beetle uses the starlight from the Milky Way as a celestial compass to navigate through the night. Of course, this means there could well be other insects using it for the same reason. But now, with light pollution, the Milky Way is being rapidly eroded, lost in the glow of artificial light. This could leave species like this intrepid beetle with no way to navigate themselves, lost in an eternal artificial haze of fake starlight.

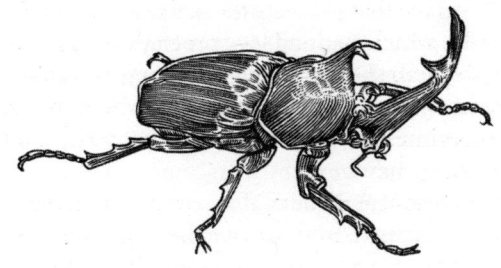

Japanese Horned Beetle

Trypoxylus dichotomus

Kabutomushi • Jangsupungdeng-i (General Beetle)

Nocturnal

Location	Average Lifespan	Average Size	IUCN Red List Status
Japan, Taiwan, China, Korean Peninsula, Vietnam, India, Thailand	4 months (as adult)	6–9cm long	Not assesssed

This is one charismatic and charming beetle from the Scarabaeidae family that has been elevated to hero status in some cultures across Asia, and it isn't hard to see why. One of the largest beetle species, the males have a broad body that is defended by deep mahogany armour. From its head protrudes a large 'antler'. This Y-shaped horn can be as long as 3 centimetres, and is used to defend the territory of its favourite tree, guarding the prized sugary sap that lies under the bark and, of course, to compete for females. If another male comes charging in like a jousting knight, they will use their horns like a lance. After sizing one another up, if they each fancy their chances at emerging victorious, they will try and flip one another onto their backs by scooping their 'horns' under the challenger. The females also have horns, but they are much smaller than those of their male counterparts. However, this doesn't stop them from charging at enemy females, attempting to banish them with a battle-ending headbutt.

This battling behaviour has earned them many nicknames across Asia. The Japanese name *kabutomushi* literally translates to 'helmet beetle'. A *kabuto* was a helmet used by ancient Japanese warriors that later became synonymous with the traditional armour worn by the samurai. These beetles have come to be revered as symbols of strength and are very popular as pets across Asia. Youngsters will go out to hunt for the beetles in the wild, coming together to make them battle one another. Some have linked this to the inspiration for the wildly popular Pokémon games and TV series, with the Pokémon Heracross bearing more than a passing resemblance to this feisty forest-dweller.

Luna Moth

Actias luna

American Moon Moth

Nocturnal

Location	Average Lifespan	Average Size	Average Weight	IUCN Red List Status
North America	1 week (as adult moth)	11.5cm	1.7-2.8g	Stable

Butterflies are revered for their beauty and grace by humans, who often relegate moths to being the butterflies dusty and dull cousin. Well, those people have never laid eyes on what must be one of the most gloriously gorgeous insects on the planet, let me introduce you to the luna moth. They are more enchanting than a Bridgerton in a ballgown, these giant silk moths are resplendent with large, whimsical wings of laurel green delicately edged with crimson velvet that dramatically scallop into an elegant swallow-tail train. At the centre of its lower wings, two bejewelled brooch-like spots capture your gaze, its upper wings are adorned with two crescent moons. A large species, they can easily fill the palm of your hand.

They are a bit of a Cinderella story, having around just 7 to 10 days to find a companion, mate, and lay their eggs before dying. In their adult moth stage, they have no mouth or digestive system and must rely on the food they bulked up on during their hungry caterpillar stage to see them through their final days of life.

Wholly nocturnal, they are of course drawn to light like moths to a flame. They are known to travel large distances in search of a mate, but are drawn to artificial light sources that can cause them to become trapped by the light until the morning comes, and likely be breakfast for the early bird. It has long been suspected that the Moon and sources of natural light play a role in the life of moths, but it was only a very recent scientific study that found moths use light as a way of navigating themselves. They can 'orientate' themselves by the natural light of the Moon, always knowing which way is up, but artificial light sources confuse these signals and have a devastating impact for our insects.

Nightingale

Luscinia megarhyncho

Filiméala • Eos • Nihtegale

Diurnal and Nocturnal

Location	Average Lifespan	Average Size	Average Weight	UK Conservation Status
Europe, sub-Saharan Africa, Asia	2 years	15–16cm long	18–24g	Red

Romanticised since the dawn of time, the nightingale has been immortalised as a roosting Romeo in poetry, literature and song for hundreds of years, inspiring literary works from *The Odyssey* to Shakespeare, and the poetry of Shelley and Keates. This cryptic bird would not normally get hearts racing – it is a relatively plain bird, a smidge bigger than a robin, pale brown and buff all over, but it is its song that has made hearts soar throughout history.

Around the world they have celebrity status and are the national bird of not one, but two countries: Ukraine and Iran.

They are a migratory species, spending their winters in sub-Saharan Africa before spreading their wings and heading to European woodlands for the start of spring. The secretive serenader will arrive and find a suitable home in which he hopes to tempt his lady love, then as the females start to arrive, it's time to showcase that vocal talent. Like many other species they migrate under cover of darkness, and this is why the males sing through the night, serenading away under the Moon and the stars. The lonely Casanova croons until he successfully charms a female, so the later into spring you hear him singing, the more unsuccessful his songs have been.

They have quite the repertoire, able to trill, whistle and warble over a thousand sounds. This concerto is astonishingly varied when compared to the range of other species like the blackbird, another famed vocalist, but with a limited range of just a hundred vocalisations.

Its name has roots in Old English and German, essentially meaning 'night singer'. They have nearly been lost to the UK, with the population collapsing between 1967–2007, where we lost 91 per cent of our most beloved bird. Their threats are many and varied, including climate change and loss of habitat.

Although not a fully nocturnal species, the night would not sound right without the melodic romance of the nightingales' song.

Indian Crested Porcupine

Hystrix indica

Sahi • Ittawa • Mullam Pandi

Nocturnal

Location	Average Lifespan	Average Size	Average Weight	IUCN Red List Status
Southern Asia and Middle East	27 years (in captivity)	70–90cm long	11–18kg	Least concern

A very large rodent, these 'Old World' porcupines have roamed continents to find homes from the heady heights of the Himalayas to the banks of the River Ganges and beyond.

They are the largest rodent in India, cloaked in a magnificent set of quills that can be 40 centimetres long. They are of a stocky set, with a rounded snout and the face of an oversized guinea pig. Their crowning glory is their quiff of quills that sits on top of their heads. It's the size of these quills that sets them apart as 'Old World' porcupines from the 'New World' species. If threatened, these are their first line of defence, raising them in warning if a predator gets too close. Around their tails, they have hollow quills that they can shake, adding to their battle growls and stomping feet to intimidate their attacker. If that doesn't work, they will lead a charge with those sharp quills that can easily pierce skin.

In Tamil, they are known as *mullam pandi*, which translates to 'prickly pig', similar to the etymology of 'porcupine', which derives from the Latin words for 'thorn pig', an apt moniker for this quilled quintessence of the night.

A bit like werewolves, they try to avoid being out around the full moon. The full moon can be exceptionally bright on a dark night, meaning if you're out looking for midnight snacks, something might spot you too! They are wholly nocturnal, and this seems to have stopped them spreading too far north in their range, as they need a minimum of 7 hours of darkness at night to comfortably go out foraging. They are known to be light shy, avoiding areas that are artificially lit at night.

Fennec Fox

Vulpes zerda

Fannak • Akori • Eresker

Crepuscular and Nocturnal

Location	Average Lifespan	Average Size	Average Weight	IUCN Red List Status
Sahara, including Morocco, Egypt, Tunisia and Sudan	10 years	36–40cm long (excluding tail)	0.1–1.4kg	Least concern

Out in the dark of the desert night, the fennec sits a while and stares to the skies. Perhaps a meteor flashes above, burning brightly, reflected in the doe eyes of the fox, or a rocket steadily makes its course through the starlight, on its way to get us closer to Mars, unbeknown to the fox whose feet will never touch the alien soil that feels so similar to the shifting sands of her home.

Daintiest of all the dog species, the fennec fox is the elfin spirit of the Sahara. Their home can be vicious, with high winds and scorching sands that are whipped into the air, turning the whole landscape into a transient tyrant that blocks out the searing sun with plumes of dust that fill every inch of the sky. But they have adapted to rule these serpentine sands, that eternally shift and play tricks on those passing through. Their most notable feature must surely be their giant ears, which stand starward like satellite dishes, extending up to 15 centimetres high. Not just for display, they are woven with blood vessels that transfer heat away from the fennec's body to keep it cool. Of course, giant ears mean supernatural hearing, they can tune in to the smallest of sounds, such as a tunnelling rodent deep under a dune or a beetle marching by starlight many metres away. They are the colour of the sands they live on, fur blending perfectly into the ocean peaks and troughs of the desert, bar the tips of their tails, dipped into the inky black of the Saharan night.

The structure of their paws has inspired designs of space rovers, as they're so good at moving through soft soil and sand. Their most impressive feat is their ability to live their entire lives without needing to find a water source as they get all the moisture they need from the things they eat.

Threats to their survival include human influences, such as hunting to sell them into the illegal pet trade and climate change which will cause a loss of 40 per cent of their habitat by 2050.

European Starling

Sturnus vulgaris

Drudwy • Druid • Druideag

Diurnal and Nocturnal

Location	Average Lifespan	Average Size	Average Weight	UK Conservation Status
Europe, Asia, Africa (Native)	3 years	19–23cm long	58–101g	Red

Feathers as black as night and speckled with bright white stars, but this species is not named for its plumage, but for the shape it makes in flight, resembling a four-pointed star. During the autumn and winter, these birds flock together en masse to create one of the most stunning spectacles in nature; the murmuration.

Thousands upon thousands of whirring wings collectively dance in the sky, an enthralling event that celebrates the end of the day and inivites the coming night to draw in.

Active in the day, these birds can also be active during the night. Like most other migratory bird species, they use the cover of darkness to take to the skies to fly to faraway shores. But how do they find their way through the dark night? They look to the heavens for answers. One of the navigation techniques used by starlings (and many others) is the natural light cues that come from an unpolluted night sky, the stars provide cues that lead the way.

They are one of a myriad of bird species being impacted by light pollution. It messes up their ability to read these natural nocturnal cues, blocking them with our artificial stars and moons that we shine up to the sky from Earth. Not only that, when they finally land, exhausted from their efforts, light pollution prevents them resting properly. Even the light of a full moon impacts how much sleep they get, so imagine what billions of bulbs are doing?

In Belfast, Northen Ireland, a much-beloved starling roost was disturbed so much by light pollution that they took to the skies and left. The Albert Bridge was home to a large roost that drew crowds from across the city to watch their aerial displays, but when lights on the bridge were converted to LEDs, the birds left, and Belfast was all the poorer for it. Thankfully, the city listened to residents and worked to reduce the lights' impact, and slowly, year on year, the starlings are returning home.

Puerto Rican Boa

Chilabothrus inornatus

Culebr • Yellow Tree Boa

Nocturnal

Location	Average Lifespan	Average Size	Average Weight	IUCN Red List Status
Puerto Rico	20 years (in captivity)	1.9m long	3kg	Least concern

A tropical island paradise in the Caribbean Sea is a good place to live if you're a snake, with lots of sunshine to warm that scaly skin and cold blood. That's why the lifestyle of this snake is a little surprising. Predominantly a nocturnal species, the boa hunts at night, and not only that, they totally shun the sun to take up residence in the chasmic caves of Puerto Rico, where the world is so dark you can't see your hand in front of your face. Cavernas del Rio Camuy is the third-largest cave network in the world, and the largest in the Western Hemisphere.

Despite being underground and hidden from the embrace of the Caribbean sun, the cave network remains warm, around 16–22°C, so it's the perfect place for this skilful serpent to, quite literally, hang out. An arboreal species, the boa can take to the branches of trees in the forests. It's adapted this skill and taken it underground, where it will stealthily stow itself into the walls and ceilings of caves, waiting for dusk to settle. Once twilight approaches, a susurration spreads through the darkness as thousands of wings activate, stretching and shivering themselves free from slumber. Antillean ghost-faced bats take to the night in their thousands, rushing the cave entrance like a winged wave. The boa has strategically spread itself along a trailing vine, its muscular brown body now perfectly positioned, coiled like a trap strung across the cave tunnel, waiting for a bat that has failed its hazard perception training. Despite being experts in echolocation, the sheer volume of bats means some fly straight into the awaiting slithering snare. The boa wastes no time, immediately twisting itself tightly around its prey, and doing what boas do best – constricting. Resistance is futile. No one escapes the vice-like grip.

Despite looking the part, these boas are non-venomous and pose no threat to humans. They tend to shy away from us but if you do try to handle one, or unexpectedly disturb it, you may get acquainted with its fangs.

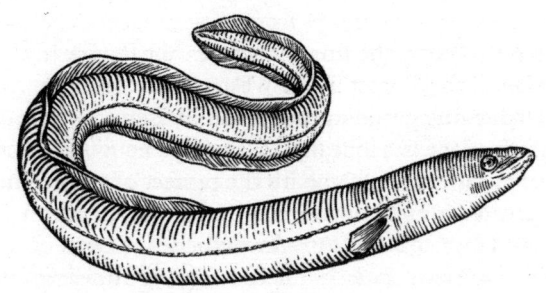

European Eel

Anguilla anguilla

Llysywen • Ael • Easgann • Eascann • Ankerias

Nocturnal

Location	Average Lifespan	Average Size	Average Weight	IUCN Red List Status
Europe, North Africa, Sargasso Sea	10–20 years	50–70cm long	1.5kg	Critically endangered

Mysterious beings that live their lives shrouded in the secrecy that the darkness of the waters offers, the European eel is a fascinating species that has been linked to European cultures for hundreds of years. At one time they were so plentiful and valued that they were used as currency. In Medieval Britain you even paid your rent in eels.

Despite being a freshwater species, they migrate a massive 6,400 kilometres from the coasts of Europe to their spawning grounds somewhere in the Sargasso Sea (although the exact location has never been found). This is one of the greatest geographical ranges of any fish. They shapeshift throughout their life cycle, starting as tiny eggs that float towards Europe on the ocean currents for a couple of years until they become larvae. Arriving at the coast, they have shifted into their 'glass eel' stage; teeny-tiny and transparent, trusting the tides to ferry them up streams and rivers before transforming into elvers as they enter freshwater; growing into yellow eels before their final stage as the silver eel, which must then find its way back to the Sargasso Sea to start the whole cycle over again.

They can live for a very long time. One eel, named Åle, became a living legend, reaching the ripe old age of 155. Back in 1859, Åle was released into a well on a Swedish farm, when it was common practice for eels to keep the well's water clean of any animals wanting to set up home there in the darkness. His death made international news in 2014. In another well, this time in Cymru, there lived prophetic eels. Lovers on pilgrimage to Ynys Llanddwyn (the Church of St Dwynwen), could stop at St Dwynwen's sacred well. Scattering breadcrumbs and laying a delicate hanky across the surface would tell if your true love would be faithful. If the eels disturbed the surface, eternal faithfulness and wedded bliss. If not? An awkward conversation on the journey home.

Since 1970, the number of eels reaching Europe has declined by 90 per cent (some estimates put the figure closer to 98 per cent). Their ability to complete their life cycle through migration has essentially been blocked by humans, with dams and weirs, river pollutants and overfishing blamed for declines. One of the biggest and newest threats? Light pollution. When artificial light spills onto waterways, it creates a barrier for eels and many other aquatic species who are highly sensitive to light. Eels are nocturnal, and this unnatural day time being thrust upon them is literally killing them.

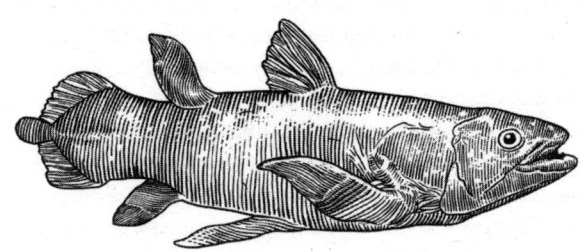

Coelacanthus

Latimeria chalumnae

Old Four Legs • Seelakant • Gombessa • Skúfur

Nocturnal

Location	Average Lifespan	Average Size	Average Weight	IUCN Red List Status
West Indian Ocean	60–100 years	1.8m long	90kg	Critically endangered

From the depths of not only the ocean, but time itself, this fish has risen like Lazarus. Only this wasn't a measly 4-day post-death resurrection, this beast came back from the dead after 70 million years of assumed extinction! They are a member of the 'Lazarus taxon', a name given to species thought to be extinct before reappearing, often dramatically, like, 'Surprise! I'm alive!'. Coelacanthus was no different, each sweep of its newly alive tail caused ripples that traversed the world's oceans. Little did this fish know as it went about its nightly hunt, that above its watery world the life of celebrity awaited.

It was found swimming around the sea off South Africa in 1938, after local fishermen hauled it in with the rest of their catch. It caught the eye of Marjorie Courtenay-Latimer, a museum curator who thought this a peculiar *pesce* and dutifully drafted up drawings to send to an ichthyologist (a person who specialises in the study of fish) who immediately sent a frantic message via cable – 'MOST IMPORTANT PRESERVE SKELETON AND GILLS = FISH DESCRIBED'.

This fish looks like it belongs in the prehistoric and somehow swam its way out of the history books into the modern day, and that's because they have. They are thought to be one of the key species that show how some life forms transitioned from an underwater world to one with solid earth beneath their finny feet, they have four fins that move like something on land with four legs would move. They have thick, armour-like skin and a huge jaw they can unhinge to consume prey. We aren't sure how long they live, but the oldest specimen ever found was at least eighty, suggesting they are long-lived beasts. Living in the dark, life takes a slower pace, from the way you breathe and move, to things like reproduction. It is possible gestation in this species takes up to 5 years, after which they give birth to live young. There is so much more to learn about life on Earth just waiting to be discovered down in the dark.

Hawaiian Spinner Dolphin

Stenella longirostris longirostris

Nai'a • Long-beaked Spinner Dolphin

Dirunal and Nocturnal

Location	Average Lifespan	Average Size	Average Weight	IUCN Red List Status
Hawaii	20 years	1.3-2.4m long	50-79kg	Least concern

All around the coasts off the islands of Hawaii swim the ocean's joy. Acrobats of the open seas, the spectacular sights of a pirouetting dolphin leaping into the air, spraying a trail of crystals and rainbows in its wake.

These dolphins are incredible aerialists, capable of leaping up to 3 metres into the air and can spin up to seven whole rotations before returning to the water once more; they can pull off aerial stunts Olympic gymnasts could only dream of. We don't know exactly why they spin, but researchers think it could be part of developing social bonds, sometimes courtship and other times just for the sheer joy of it. Once one dolphin starts, others will join in with the displays, so it seems there is some truth to it being a social or playful activity. They are a very social species and can live in huge pods comprising over a thousand individuals.

They are a slightly strange dolphin, in that they have a day and night routine. They mainly spend their days just off the coast, relaxing and sleeping close to the surface of the water, charging themselves up for showtime come sundown. After sunset, they head off to forage and fish, hunting throughout the night in a scattergun approach. Taking advantage of the sea after dark is a smart strategy, as each evening when the Sun sets, there is a mass migration as tasty things that live in the darker parts of the ocean rush higher up the water column to take advantage of the newly darkened space. This way, the dolphins can take their pick of a myriad of species. Due to their nocturnal lifestyle, it is imperative these dolphins are allowed to rest during the day, close to shore where they're safe from predators. Unfortunately, this makes them easy to find for people who want to spot dolphins, with boat-dwelling tourists disturbing them from their resting places in order to swim with them.

Whenever you visit the ocean, remember to respect the water and everything in it. You're entering their world and humans need to leave our wonderful watery wildlife alone.

Halloween Hermit Crab

Ciliopagurus strigatus

Unauna • Cone Shell Hermit Crab

Nocturnal

Location	Average Lifespan	Average Size	IUCN Red List Status
Indo-Pacific Ocean	10 years	6cm long	Data deficient

Should you take a trip to the beach after sunset you may be treated to the sight of a nocturnal trickster scuttling along the coral reefs and shallow seas of tropical waters. A pretty coloured shell seems to be moving before your very eyes, turning in the sand. Then you spot one leg, then another, sticking out from under the shell and, before you know it, off it goes, scampering over the soft, sandy shores.

Trick or treat! It's a Halloween hermit crab. This species is among the most brightly coloured of the hermits. It gets its common name thanks to its colours being reminiscent of Halloween decorations, with its bright pumpkin-orange and blood-red banded legs, and under its shell is a ghostly white body, protected by the remains of some other poor soul.

Hermits are scavengers and get to work cleaning up all the dead stuff that exists on the seabed, and in a move that may seem ghoulish to some, it's always on the lookout to upgrade its mobile home; the shell that it carries on its back. It acts as a hiding place from which it can startle passersby or use for protection from bigger, scarier beings. It's not always easy, finding the perfect home, it's not like they have Zoopla under the sea. They'll never be chain-free buyers, either, with every moult the crab must find a bigger shell, which can lead to a chain of shell swapping by hermits. Once a hermit has found a larger, vacated shell, it can trigger a swap shop frenzy, with hermits all in the market to size up as each shell gets passed down to the next crab. Sadly, there's been frequent reports of hermits using litter or plastic items for their new dwellings.

Hermit crab species occur all over the world, giving you the perfect excuse for an evening stroll on the seashore to try and spot one of these fascinating creatures.

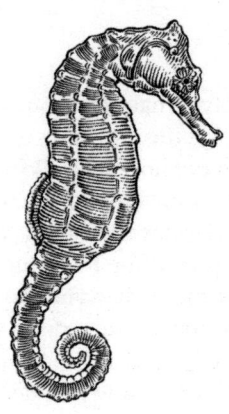

Pacific Seahorse

Hippocampus ingens

Caballito de mar • Caballito del Pacífico

Nocturnal

Location	Average Lifespan	Average Size	Average Weight	IUCN Red List Status
Eastern Pacific Ocean	3–5 years	20–30cm long	200g	Vulnerable

Dragonlike in appearance and dancing their way through the ocean are these miniature marine marvels. Their scientific name translates to 'horse sea monster', which belies the true beauty that these minibeasts hold. Their horse heads could be carved from jewelled corals, and their curled, serpentine tails wouldn't go amiss in a fantasy realm.

Found in a variety of habitats from mangroves to seagrass meadows, Pacific seahorses are one of the largest of their species and are nocturnal. With a prehensile tail, they can anchor themselves on kelp, coral or blades of seagrass while hunting for their food. They don't have stomachs, so must pretty much be eternally grazing, eating up to thirty times a day. They have skin instead of scales, which can come in a variety of colours, from maroon to canary yellow; as colourful as a chameleon, seahorses can also move each eye independently. They are not the strongest of swimmers given their small size. The fin on its back beats up to fifty times a second, a hummingbird of the sea, but they are at the whim of the ebb and flow of the ocean's tides and waves. Being this fragile in something as big and brutal as the ocean can mean these tiny treasures can become fatally exhausted by rough seas.

The seahorse is probably most famed for the fact that the males birth the young, fertilising the young in a brood pouch and brooding up to 2,000 young at one time! The poor bloke must then labour each tiny seahorse – disappointingly called 'fry', not 'foals' – one by one. It is an exhausting test of endurance that can last hours.

Around 20 million seahorses a year are sold into the aquarium trade, dried as grim curiosities, or shipped to China for use in medicinal markets, leaving this species extremely vulnerable to extinction, especially combined with other factors like pollution and climate change.

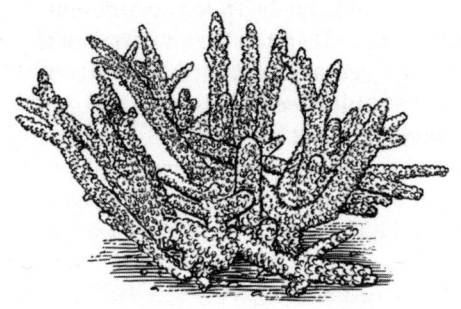

Staghorn Coral

Acropora cervicornis

Corail corne de cerf

Nocturnal

Location	Average Lifespan	Average Size	IUCN Red List Status
Bahamas, Caribbean and Florida	100+ years	1.5–2m high; 10m wide	Critically endangered

Coral is an incredibly diverse and fascinating being. Commonly mistaken for plants, these organisms are actually a type of animal. Made from a collection of thousands of tiny animals called polyps, and protected by a hard skeleton, they come in a dazzling array of colours, shapes and sizes.

Staghorn coral is made up of 160 different species, all playing a crucial role in the health of one of the planet's most important ecosystems. It can take 10,000 years to form into a complex reef habitat that hosts an array of life in its branches.

This species used to be the most abundant in the Caribbean, however it was decimated when a disease spread through the colony and left only 3 per cent of the colony surviving. It is a golden-brown colour and grows into branches that resemble the horns of a stag, hence the name. It is one of the ocean's fastest growing coral species and can increase in growth up to 20 centimetres a year under the right conditions.

Like many species, this coral is a night hunter. As darkness tiptoes across the reef this coral unfurls a secret weapon – miniscule tentacles that pack a sting when catching its prey of tiny plankton that float around the sea. Their life cycle is linked closely to the natural ebb and flow of the lunar cycle as well as the tides.

Staghorn are hermaphrodites. They release both egg and sperm in what is called 'broadcast spawning'. Once a year, after the full moon of August or September, they set loose their pre-progeny pieces in the hope they will go on to form new colonies.

Our corals face a latitude of threats. Climate change causes sea levels and temperatures to rise and increases the number and strength of hurricanes, meaning these delicate ecosystems desperately cling to existence.

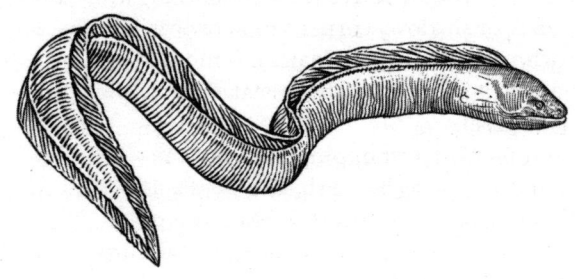

Giant Moray Eel

Gymnothorax javanicus

Almang • Dabea • Puhi • Maoa'e • Tiohu

Nocturnal

Location	Average Lifespan	Average Size	Average Weight	IUCN Red List Status
Indo-Pacific region	30 years	3m long	30kg	Least concern

The coral reef seems like sunshine itself has erupted under the water, from its carnival of colours to its riotous cast of marine life. By daylight, it is a sensory overload of wonder but don't be fooled into thinking this party stops when the Sun goes down, for when the lights go out, that's when the nightlife really kicks off.

Found in tropical lagoons and coral reefs, this nocturnal predator shirks the sun, slinking away into the safety of shadowy corners and crevices. This is one grumpy-looking fish, dark and mottled brown in colouring, it could easily be mistaken for being a part of the rocky reef.

Coming in at a whopping 3 metres, this eel is truly elephantine. It may be a fish, but these eels don't have scales. Instead, their skin is covered in a protective mucus layer. They have a very long dorsal fin that starts at the back of their head and travels all the way down to the tip of their serpentine tails.

It roams the reef, combing it thoroughly for crustaceans, but this hunter doesn't work alone. He works in a gang, often being joined by the company of another nocturnal specialist, the roving coral grouper. Together they work in tandem to ransack the reef: the eel's shadowy presence scaring the fish, upwards, into the waiting jaws of the grouper. A famous pair of moray eels, Flotsam and Jetsam, skulked across the silver screen as familiars of Ursula, the sea witch, in Disney's *The Little Mermaid*.

Take a closer look at this enormous eel and you will see a jagged jaw, lined with razors. As he sits with his mouth wide open, a silver flash can be seen flitting between his fangs. That is the brave little dentist of the reef, the bluestreak cleaner wrasse, working away to polish the teeth of these formidable fish. Speaking of jaws, this creature has a set of two. The 'pharyngeal jaw' sits behind the front jaw, and during feeding it will thrust this secondary jaw forward to clamp down on its prey and pull it further into its throat to ensure not a morsel slips away.

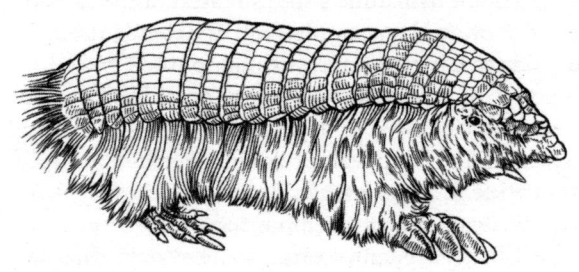

Pink Fairy Armadillo

Chlamyphorus truncatus

Pichiciego

Nocturnal

Location	Average Lifespan	Average Size	Average Weight	IUCN Red List Status
Argentina	5–10 years	7–8cm long	120g	Data deficient

Despite looking and sounding like this being is a thing of make-believe, I promise you the pink fairy armadillo is a living, breathing creature that has shuffled straight from the pages of a fabulous fairytale to roam under the twinkling lights of the stars and the deep darkness of the underground.

Pale, Barbie pink, and the right size for a Polly Pocket, this miniscule armadillo is the smallest of the armadillo species. Covered by a supple shell, it defends itself the same way as larger species, curling up into a ball to defend its delicate underbelly from predators. Its body appears to be covered in downy dandelion fluff, giving this armadillo the appearance of a hamster who decided to try on some battle armour. It has oversized feet with strong claws, like that of a mole's, and powerful forearms, making them excellent excavators. They're known for their ability to quickly 'sink' into sand, giving way to the nickname 'sand swimmers', as they can quickly shovel their way to the safety of the subterrain.

They are strictly nocturnal and given that they spend their lives mainly in burrows, they have small eyes without the best vision. Instead, they use smell and touch to find their way through the darkness to feast on favourites, such as ants and other bugs.

They are elusive and rarely seen, even by those trying to study them, with some individuals never spotting a single armadillo despite working in their habitat for over a decade. We know they are at risk from loss of habitat and climate change, as this species is very sensitive to even small changes in its environment such as fluctuations in temperature or increased rainfall, which can flood burrows and cause them to evacuate. Domestic dogs also prove to be a big issue for this small species. More nature-friendly land management needs to be adopted to ensure this mystical being isn't relegated to actual fairy tales for good.

Pacarana

Dinomys branickii

Branik's Rat • Terrible Mouse • Guagua loba

Nocturnal

Location	Average Lifespan	Average Size	Average Weight	IUCN Red List Status
Western South America	10 years	79cm long (excluding tail)	15kg	Least concern

You may have heard of Supermouse, but have you heard of 'terrible mouse'? This rarely seen and robust rodent lives in the cloud forests of the Andes and the Amazon basin, where it rummages around in the darkness for food. It leads an unhurried life and is slow moving, taking its time to find the finest fruits and leaves for its evening meal.

They look like a guinea pig, rotund barrel bodies and large heads, with a set of long and wiry whiskers. Covered in dark fur, they are speckled with a constellation of white markings that run down its flanks, ending in a bushy tail.

Its name means 'fake paca', the paca being another animal, which although unrelated, it does look similar to. The scientific name of its family, Dinomyidae, means 'terrible mouse', like 'dinosaur' means 'terrible lizard'. Bit harsh. There is nothing terrible about these placid beings.

Rarely seen in the wild due to being wholly nocturnal, most of what we understand comes from studying those that live in captivity, where they don't always act as they would in the wild. They have surprisingly shown a curiosity in humans, even approaching them to interact with keepers – much like your pet cat when vying for a chin rub – suggesting that they are vulnerable to being exploited by humans, especially as they are slow moving and easy to capture.

They foot stomp as a form of communication, much like rabbits, and will sit back on their hind legs when dining on leaves and shoots, which they grasp in their front paws like a hefty squirrel eating a nut. The males like to romance their females, and croon to woo them, singing for up to 2 minutes to hopefully sweep a female off her feet.

Officially listed as Least Concern, little is known about the lifestyle of this animal, including how many exist in the wild or how long they really live for, meaning this mammal remains a mystery for now.

Striped Hyena

Hyaena hyaena

Tzebua • Zevoa

Nocturnal

Location	Average Lifespan	Average Size	Average Weight	IUCN Red List Status
Sub Saharan Africa, Middle East and Central Asia	10-15 years	90-130cm long	30-55kg	Near threatened

The hyena's coat may be striped, but his past is definitely checkered. This creature has been haunted throughout history by malevolent myths and sinister superstitions that have led to it being feared and terrorised by humans. Its cardinal sin? Being a creature of the night (and looking a bit weird).

Much of the mystery that shrouds the hyena revolve around it being able to transform from animal to human. They do cut an odd figure, with a bulky body on a set of legs that are much shorter at the back than the front, giving a sloped spine and a scurrying, surreptitious gait. They have the look of a startled dog caught in the act of trying on the suit of a zebra. The striped hyena is the smallest of the species and doesn't 'laugh' like the more commonly known spotted hyena.

They are scavengers and do a very good job of clearing up dead stuff. It is perhaps this habit that led to sinister tales of vampiric hyenas feasting on the blood of humans. In some cultures, they are depicted as being in cahoots with witches, supposedly riding on their back to find human bodies as the witches hunted for souls of the deceased. This seems rather unlikely for an animal known to feign death when attacked.

They are a shy animal and don't tend to live in packs like the very social spotted hyena, and instead prefer to live alone or in pairs. This, and their nocturnal ways, means they are very seldom seen by humans. They are the national animal of Lebanon but despite this are still very much unwelcome in most places they exist and are often persecuted. In reality, they just want to be left to their shadowy selves.

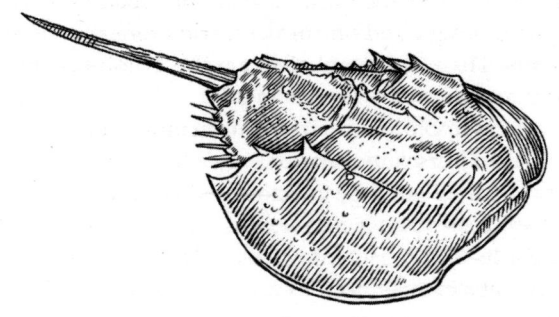

Atlantic Horseshoe Crab

Limulus polyphemu

Horsefoot • Saucepan • Helmet Crab • Dolkhalke • Mex

Nocturnal

Location	Average Lifespan	Average Size	Average Weight	IUCN Red List Status
Atlantic Coast of North America	40 years	38–48cm long (excluding tail)	4.8kg	Vulnerable

The creature that laughed in the face of extinction, not just once, but three times. The horseshoe crab is an extraordinary example of evolutionary prowess and, once it found its winning strategy, has remained pretty much unchanged for millions of years. They appear in the fossil record over 400 million years ago, making them a living dinosaur and one of the world's oldest natural wonders. Their closest relatives are the Trilobites, long extinct and now only seen as fossils.

They resemble an oversized and armoured tadpole, with a hard horseshoe-shaped exoskeleton (or shell) and a long, pointed tail. The tails look lethal, akin to a ray, but they're completely harmless and are mainly used for righting itself if it gets flipped over. Despite the name, they're not a crab, or even a crustacean. They are most closely related to spiders and scorpions.

Horseshoe crabs have been mooching about our oceans long before we started installing artificial lights here, there and everywhere. They evolved to respond to the cyclical rhythms of the Moon and the tides. They are nocturnal and have ten eyes that are incredibly sensitive to their environment. They swarm the beaches in their thousands to mate and lay eggs, but with our use of light we are upsetting the balance of a life that existed peacefully for a span of time we can barely contemplate and may be affecting their mating schedules.

Despite surviving the extinction-level event that saw off the dinosaurs, these amazing creatures might not get to outlive us. Two species are classed as under threat of extinction due to human activity, such as overharvesting for use in the medical industry, climate change, and degradation of habitat.

Giant Pangolin

Smutsia gigantea

Scaly Anteater • Letermago • Haka • Inkakha • Isambane • Kakakuona

Nocturnal

Location	Average Lifespan	Average Size	Average Weight	IUCN Red List Status
Central Africa	20 years (in captivity)	136-180cm long (including tail)	33kg	Endangered

Resembling an animated pine cone, this creature looks like it tumbled from the wildest of imaginations. They really have to be seen to be believed, and even then you will be pinching yourself to check it's not a dream. These unassuming and retiring animals walk on their hind legs, stooped over with their front claws clasped in front of them, like an elderly gent making his way to the shop for his morning newspaper. Their long tails are held out behind them for balance, a bit like a walking stick. David Attenborough stated that they are 'one of the most endearing animals I have ever met', and you know Sir Attenborough tells no lies.

Pangolin comes from the Malay word for roller, *pengguling*, an apt description of the defensive stance this creature takes when threatened, rolling itself up into an armoured ball. They are covered in scales made of keratin. This quirk makes them the only mammal species covered in scales from tip to toe. They are predominantly nocturnal and solitary as a species, preferring to live out a quiet existence. When a baby is born, the adorably named pangopup will hitch a ride on its mother's tail until it is strong enough to break into termite mounds itself.

Their tongues are an outrageously long 70 centimetre tool of stickiness, which they use to lick up termites and ants after breaking into their mounds by using their sharp claws and weight to collapse the structures. They don't have any teeth, so they eat rocks to aid the digestion of their food in their stomachs. Being toothless doesn't trouble them, they eat around 70 million insects a year! They can seal their ears and nostrils shut to stop any insects finding their way into places you don't really want them to venture.

Tragically, pangolins are thought to be one of the world's most trafficked animals, exploited for their scales and meat. All eight pangolin species are at risk of extinction. We know very little about them due to their nocturnal and secretive habitats, so for now, they remain a cryptic creature of the night.

Common Leopard Gecko

Eublepharis macularius

Kkhan-kkhain • Korrh Kirly

Crepuscular and Nocturnal

Location	Average Lifespan	Average Size	Average Weight	IUCN Red List Status
Afghanistan, Iran, Pakistan, India and Nepal	5 years	20-28cm long (including tail)	70-80g	Least concern

In the wild, this gorgeous gecko can be found crawling across the rocky grasslands of South Asia, where it will spend its day sheltering in nooks and crannies, safe from the prying eyes of predators.

Named for their pretty spotted patterns, these geckos tend to be the colour of golden sand, freckled with brown leopard spots. They even crawl like a leopard, with their bodies slung low to the floor, and although they can climb, they lack the sticky toe pads of other species so can't climb vertical surfaces. Instead, they have claws at the end of their toes that help them grip and scramble over rocks and branches. They have large green eyes set either side of their wide heads and are one of the only gecko species with eyelids, giving them more expressive faces than other geckos. They also have excellent night vision and can even see colours in the dark!

Their tails are crucial to their survival for a number of reasons, a bit like a multitool it has a multitude of uses. They are long and can grow to be quite chunky, which they handily use to store fat. This means when times are hard and food is scarce, they can access this fat store to keep themselves going. Not only that, but their tail can be used to make a quick getaway. If they find themselves in a sticky situation, they can detach their tails and buy themselves some time to make a sneaky escape. Amazingly, they can regrow their tails in as little as a month.

This species is naturally quite placid, and paired with their pretty faces, has made them popular as pets. Despite needing to be fed live insects, they have found their way into many homes and hearts. In captivity they can live for 20 years or more, so are a huge commitment. However, they are not always respected as nocturnal creatures, which can lead to them being unhappy.

Little Blue Penguin

Eudyptula minor

Kororā • Fairy Penguin • Little Penguin

Diurnal and Nocturnal

Location	Average Lifespan	Average Size	Average Weight	IUCN Red List Status
New Zealand and South Australia	6-8 years	30cm tall	1kg	Least concern

As the name might suggest, the little blue penguin is the smallest of the penguin species, not much taller than a ukulele or a computer keyboard. They are a gorgeous slate blue along their backs with bright white bellies. Quite the beauty of the penguin word, their eyes are a dazzling azure. They have pale pink feet that waddle this penguin through the darkness across the islands they call home and into their burrows. In Australia they are commonly known as 'fairy penguins' due to their small size and enchanting characters. Their scientific name means 'good little diver'.

They are a noisy neighbour and their vocal displays can be heard throughout the night along the shores of the islands that they call home. By day these birds raft out at sea, where they will forage for almost anything they can fit in their beaks like jellyfish. Each night, you can watch the spectacle for yourself as 40,000 of these petite penguins parade from the sea to the shore of Australia's Phillip Island.

Diurnal and nocturnal, they are wholly nocturnal when on land as a response to predators. This also means that they are sensitive to artificial light sources, which can affect their behaviours. Being this small means you are incredibly vulnerable to attacks or becoming an hors d'oeuvre. Due to introductions of predators like foxes, dogs and cats, one island found its colony almost wiped out when numbers dwindled to below ten individuals in 2005.

Luckily an unlikely saviour was found in Oddball, a Maremmano sheepdog. This breed, which looks like a golden retriever and a polar bear had babies, is typically used for the livestock protection of your usual farmland animals. No one had ever used them to protect wildlife before. Thankfully, Oddball took her new job very seriously, and successfully saved the Middle Island colony and blazed a trail for the creation of the Penguin-Protection Project. Dogs are not just a man's best friend, it seems!

Coyote

Canis latrans

American Jackal • Prairie Wolf • Coyotl

Cathemeral and Nocturnal

Location	Average Lifespan	Average Size	Average Weight	IUCN Red List Status
North and Central America	6-8 years	1-1.3m long	15-20kg	Least concern

The wily coyote embodies both the fox and the wolf, with its sharp senses, pointed features, bushy tail, and howling call. It has been a staple of the American wilderness since ancient times, weaving its way from cosmology into cartoons.

They get their name from the Nahuatl (Aztec) word *coyotl*, which was then Hispanicised into 'coyote'. They have been revered and feared in equal measure across the ages, but they were especially respected in Mesoamerican cosmology. In Aztec beliefs, the god Huēhuecoyōtl (meaning 'very old coyote') ruled over music, dance, and mischief. Often depicted as a dancing coyote with human hands and feet, he was renowned as a trickster and could even change gender. In the folklore of North American Indigenous people, the coyote is a character that weaves its way across many of the different cultures, often with divine powers, such as one myth in which a coyote saves the world from an eternal winter caused by the havoc of ten evil moons.

Highly adaptable, their range has had to spread in response to the human impact on their natural habitat. They no longer stick to prairies and deserts, and are now blazing a trail into some of America's biggest cities. Coyote have become more nocturnal in urban areas where whole communities exist, rarely seen by humans, as they do their best to avoid us as much as possible. In places like Chicago and New York, researchers have studied coyotes that have shown the ability to follow traffic signals, waiting at lights until it is safe to cross the roads.

They are very fast, easily hitting speeds of 40mph when on the hunt for rodents or hares, meaning they can sprint at twice the speed of their cartoon nemesis – the Roadrunner.

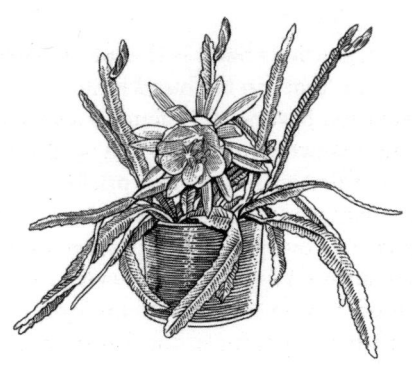

Queen of the Night

Epiphyllum oxypetalum

**Princess of the Night • Dutchman's Pipe Cactus • Tan Hua
• Orchid Cactus**

Night Flowering

Location	Average Lifespan	Average Size	IUCN Red List Status
South America	10 years	5-6m long	Least concern

The night is full of hidden wonders that mainly go
unseen by the human eye, as we find ourselves tied to the
relentless rhythms of the rising and setting of the Sun. A
mysterious world comes alive when we close our eyes, an
intricate and delicate dancing of species that care not if
they are witnessed by you and me.

One such mysterious habit is that of the queen of
the night, a plant that can be found in its native ranges
across Mexico and beyond. It is actually a species of
cactus, but not like you know it. When we think of cacti
our minds are filled with towering, prickly protrusions
that stand defiant under the sweltering sun. This species,
however, is far more secretive and demure in its habits.

Unlike its sun-seeking cousins, this species seeks out
shade and humidity where it will delphically drape itself
across the shadowy margins of the forest's branches. As
an epiphytic species, it grows on the surface of other
plants. Its stems are wide and flat, misleading you into
believing its stems are leaves but, just like other cacti,
it is leafless. Then, once a year, a large pearl-coloured
bud appears, locked tightly shut by winding fingers that
conceal the precious promise of a spectacle.

In a truly dramatic fashion, for one night only, this
Cinderella becomes a queen, and is certainly the belle of
the ball. As darkness falls, the clasp starts to slip and bit
by bit the jewel in this queen's crown is revealed under
the starlit sky of a summer evening. The deep darkness
of the forest is fleetingly illuminated by the unfurling
of pearlescent petals, akin to the delicate design of the
water lily, a huge bloom that would more than cover the
palm of your hand. But by sunrise, her carriage awaits,
and before the Sun can climb high into the sky she is
whisked away, for another year.

Golden Hamster

Mesocricetus auratus

Syrian Hamster • Mister Saddlebags • Abu Jirab

Nocturnal

Location	Average Lifespan	Average Size	Average Weight	IUCN Red List Status
Syria, Turkey	1.5–2 years	18cm long (including tail)	100–150g	Endangered

Small, cuddly and cute, the golden hamster is a pocket-sized rodent that can be found in arid areas of Syria and Turkey, where they were once widespread across sand dunes and deserts, living in burrows and emerging at sundown to go and fill up their larder. In the wild, they can roam up to 8 miles in a single evening, shoving their spoils into those expandable saddle bags they call cheek pouches before taking it back to be stored in their burrow. Weirdly, we take their name from the German word *hamstern* meaning 'to hoard', which is what these little critters do best.

A single hamster and her litter of pups were captured in Aleppo, Syria, in 1930 and sent off to Britain the following year. Over time their descendants got into the hands of private breeders, and a study found that all domestic golden hamsters are descended from that imported female almost a hundred years ago.

Since then, they have exploded in popularity for some inconceivable reason as a must-have pet for children. Countless hamsters have found their way into cages in homes across the world, usually in the bedroom of aforementioned child, where the hamster will do its utmost to turn your child as nocturnal as they are. Many of us have fond memories of sleepless nights as the pet hamster waddled from its cotton-ball nest around midnight, kicking off its evening by limbering up with some noisy stretches around the cage, before starting its night shift on the inexplicably always squeaky hamster wheel where it would run until dawn.

Now listed as endangered due to loss of habitat, wild hamsters in Europe have disappeared from 75 per cent of its habitat, making a case for it as the 'fastest declining Eurasian mammal'.

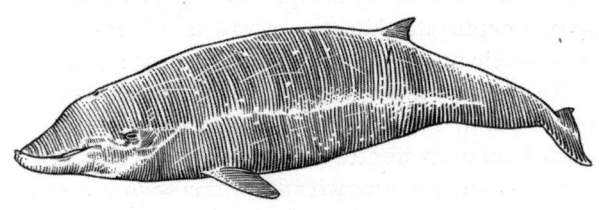

Cuvier's Beaked Whale

Ziphius cavirostris

Goose Beaked Whale • Skugganefja • Muc-bhiorach chuvier
• Blekhodenebbhval

Cathemeral

Location	Average Lifespan	Average Size	Average Weight	IUCN Red List Status
Globally across all oceans	30–36 years	5.5–7m long	3,000kg	Least concern

These beaked whales are record-breakers as the deepest diving whale species, and second deepest diving mammal species. They intrepidly explore depths where humans could only dream to go, determinedly defying physics to go and hunt where the pressure would crush all air and shape from a human. It is safer and easier to send humans to space than it is to the bottom of the deep blue sea. These whales put Jacques Cousteau to shame with their accomplishments, regularly diving down to depths of 2 kilometres, which they don with only a single breath of air. They will surface momentarily to take air deep into their lungs, before plummeting, further and further to the darkness of the ocean's midnight zone. A single breath can keep them going for up to an hour, even in such an extreme environment.

Due to their tendency to dwell at depth they are a very difficult species to study, but marine biologists love a challenge. Using satellite tags, researchers were astounded when they recorded a dive of 2,992 metres, and again when a single dive lasted for 2 hours and 17 minutes, smashing the known record for any mammal. To put that into perspective, the longest human free dive was a mere 24 minutes and 37 seconds.

Built to withstand an extreme environment, they are grey all over with a pale and bulbous melon (melon being the mass of tissue on a whale's forehead, not the fruit) that tapers into a powerful, streamlined torso, and a tail that propels them to such astounding depths where they can hunt for squid and other creatures dwelling in the darkness of our oceans.

They do not abide to being fully nocturnal nor diurnal and are active both day and night. The melon on their forehead is a crucial component of how whales communicate and navigate, it is a highly sensitive piece of equipment that helps these whales hunt in places where light ceases to exist. Whale song is so emotive, humans even sent their songs to outer space on the *Voyager* 1 and 2 spacecrafts. But the ocean is becoming a noisy place, thanks to human activity, unfortunately often leading to mass strandings of this species of whale who appear to be very sensitive to underwater disturbances.

Hoffman's Two-toed Sloth

Choloepus hoffmanni

Unau • Perezoso

Nocturnal and Cathemeral

Location	Average Lifespan	Average Size	Average Weight	IUCN Red List Status
Central and South America	20 years	58–70cm long	4.5–9kg	Least concern

Living a very chilled life hanging around the forests of South and Central America is the Hoffman's two-toed sloth. They are bizarre-looking creatures, a little like a stretched-out Chewbacca with lanky limbs and long, crescent-shaped claws that can grow up to 10 centimetres long. Given their length, walking on the ground is very hard to do so they spend as much of their lives as possible in the trees. Their seemingly eternal smile has made them a firm favourite and, possibly, the world's most famous sleepy-head.

With a top speed of 1.5mph when in a hurry, and a typical speed of 0.5mph, even the forests have tried to claim them, with green sloths not being uncommon due to algae growing in their fur. In fact, they can carry an entire tiny ecosystem on their bodies, from insects to fungi that all take advantage of the sloth's shaggy coat. Some moth species are almost entirely dependent on sloths for their entire life cycle, so the moth and the sloth live in perfect harmony. Being slow moving and living among the shrubbery means the algae also acts as much-needed camouflage for the sloths when resting on branches, very handy when you need to sleep for 20 hours a day!

Primarily folivores, feasting on leaves and tree sap, like everything else they do, their digestion is done at, well, a sloth's pace and can take weeks to break down. They are nocturnal and solitary souls. Their eyes are rare for mammals, in that they completely lack the cone cells required to see in colour, which means they are totally colour-blind and are completely blind in the presence of bright lights and daylight. Instead, they use their excellent sense of smell to forage for their favourites.

Despite being rubbish on terra firma, they are remarkably good swimmers and have been recorded covering large distances in a very short time. They are incredibly strong, too, with a grip strength three times stronger than the average human!

Javan Slow Loris

Nycticebus javanicus

Little Firefaces

Nocturnal

Location	Average Lifespan	Average Size	Average Weight	IUCN Red List Status
Java, Indonesia	20 years	29cm long	560–685g	Critically endangered

You will find this cute curiosity in the bamboo and mangrove forests of Java, creeping up the trunks of trees at a glacial pace, painstakingly placing one palm after the other, long limbs stretching out to take them to the heady heights of the treetops. They have ginormous eyes, looking like bewildered cartoon kittens in fluffy coats. They have pale, silvery-grey fur with dark facial markings, resulting in a pretty diamond shape on their forehead and a long dark stripe down their spines.

The movements of these mini mammals are stealthy and calculated as a form of protection against predators. They even have an added vertebrae compared to other primates, giving them a serpentine slink when on the move.

They may be adorable, but they have a surprising trick up their sleeves! Well, armpit. Slow loris are the world's only venomous primate. They have a special gland on their bodies that secretes an oily substance that they lick and coat their incisors with, before delivering a nasty and bone-crunching bite to would-be attackers. This isn't just a bee's sting level of venom, it can cause death in humans! Yet still these creatures are being illegally sold into the wildlife trade as exotic pets. When it's time to go foraging, a mother will slather this saliva concoction all over her baby, and park it for the night. It is thought this toxic coating deters predators and keeps the baby safe, as mum isn't exactly in a hurry.

It seems their slithery movements, clever markings and venomous bite may all come from being a bit of a copycat. When you move at the rate paint dries, it's perhaps handy to bear a passing resemblance to a lethal snake, like a cobra.

They are wholly nocturnal, their huge globular eyes adapted for darkness with a special membrane to help see at night. This membrane reflects lights back, like cats' eyes, leading to the Javan nickname of 'Little Fire Faces', as their eyes illuminate like two flames in the forest.

Grey Wolf

Canis lupus

Blaidd • Madadh-allaidh • Varg • Ulv • Lobo

Nocturnal and Crepuscular

Location	Average Lifespan	Average Size	Average Weight	IUCN Red List Status
North America. Europe, Asia, Russia	8-13 years	1.2-1.8m long	38kg	Least concern

In a snow-capped forest the frozen trees creak in the gentle but shivering breeze. The sky is crystal clear as darkness claims its place, the neon glow of a nearly new moon perforates the scenery and sets the fresh white snow to shimmering crystals under its light. A padding of paws through powder is the only suggestion a spectre haunts these hills, until a sky-shattering sound threatens to splinter the Moon to pieces. The essence of the wild. The howl of the grey wolf.

Our entire lives as humans have been intertwined with the history of this canine, from our life in the caves to the creation of Rome, they have been our shadowy companions since the beginning of time. Living in tight-knit packs, they can cover huge territories up to a thousand square miles. They were once upon a time, our most widely distributed mammal, but were nearly relegated to the story books after human persecution saw them wiped out in many countries, thanks a lot, Little Red Riding Hood!

They are active from dusk till dawn and can hunt through the night. Many wolf populations have become more nocturnal as a response to living in close proximity to humans, preferring to give us a wide berth wherever possible.

It is their chilling howls that have cemented their storybook status in the human psyche. The purpose of it is to communicate with one other, to cement pack bonds, or act as a warning to rival packs to know their place. Sadly, they don't howl at the Moon. This myth has simply come about from humans being far more likely to hear wolves howling at night, when they are most active, and is another fairytale to add to the bookshelf.

Goliath Birdeater

Theraphosa blondi

Arañas pollito

Nocturnal

Location	Average Lifespan	Average Size	Average Weight	IUCN Red List Status
Northern South America	10–20 years	12cm body; 28cm leg span	170g	Unlisted

Rainforest royalty comes in various guises, from big cats to colourful birds, but many overlook the kingdom of the creepy-crawly. This kingly specimen is worthy of many noble titles, as the largest spider on the planet by mass, this terrific tarantula is certainly the talk of the court as it creeps from its burrow to take in the evening air.

Goliath by name, gargantuan by nature, you certainly wouldn't want to accidentally disturb this sleeping giant. Despite the name, they don't frequently feast on feathers, being much more partial to mice, frogs and lizards.

Armed with 2.5 centimetre fangs, these are shown off when this beast rears up when threatened. Not only are they capable of piercing, they also can inject venom into its victims – you'll be happy to learn it is non-lethal to humans. If flashing the fangs isn't acting as enough of a deterrent, they have another secret weapon: 'urticating bristles'. These bristly barbs can be fired in defence at predators, lodging into the eyes of anything foolish enough to try and take this tarantula as a tasty treat. Though the fangs are fearsome, they have no other teeth with which to chew their food. They instead essentially spew stomach acid, or digestive juices, onto their unlucky victim. This dissolves them into a spidery protein shake to slurp from the forest floor, leaving behind the bones and skin.

If you have displeased this king, you'll know about it. They can hiss when irritated. By rubbing together special hairs on its legs in an action called 'stridulation', a technique also used by grasshoppers and crickets. If you do happen across this royal, she's more likely a queen than a king, with females outliving the males, surviving over 20 years to the male's six.

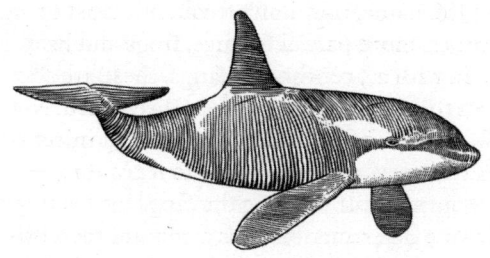

Orca

Orcinus orca

Killer Whale • Ballena Asesina • Ska'na • Kakaw'in • Qw'e Lh'ol'
Me Chen • Háhyrningur • Hvalhund • Beowas • Tsingy

Cathemeral

Location	Average Lifespan	Average Size	Average Weight	IUCN Red List Status
Globally across all oceans	40–50 years	7–8.2m long	3–5,500 kg	Data deficient but Endangered in US and Canada

Iconic and unmistakable, these are the undisputed pirate kings of the high seas. Commonly known as 'killer whales', they are actually the largest member of the dolphin family. They are incredibly powerful, their slick black and white bodies being able to torpedo through the water at speeds of over 30mph and the ability to tackle whales twice their size for food. They are instantly recognisable by their sail of a dorsal fin, it is the largest of any marine mammal, standing almost 2 metres proud of the water's surface, announcing their arrival like the Jolly Roger flying on a pirate ship.

They are the second-most widespread mammal after humans, reflected by the myriad names it has in countless languages. Their scientific name leans into their daunting distinction as skilled hunters, with *Orcinus* meaning 'belonging to the kingdom of the dead'. As they appear from the dark depths of the ocean to hunt, you can understand where the name originated from.

Not fully nocturnal, nor fully diurnal, orca are cathemeral. Refusing to obey the pull of the Sun or the Moon, they are active when they choose to be. This gives them a great advantage as they are able to hunt whenever their prey is active. Air-breathing mammals have quite a conundrum when it comes to sleeping at sea. If they slept like humans, falling into a state of unconsciousness, they would drown. To remedy this, they have what is called unihemispheric sleep, where one half of the brain sleeps at a time. The other half remains conscious, while the orca swim slowly close to the surface with one eye open for threats.

Hugely intelligent, they are a highly emotional and social species with complex brains. Each pod or orca has a unique language, like a dialect, enabling them to keep in close contact during their massive oceanic voyages. They use echolocation to successfully hunt during the day or night, rightfully earning their reputation as a formidable apex predator.

Sumatran Tiger

Panthera tigris sondaica

Macan • Maung • Harimau

Nocturnal and Crepuscular

Location	Average Lifespan	Average Size	Average Weight	IUCN Red List Status
Sumatra, Indonesia	12-15 years	2.2-2.5m long (excluding tail)	110-140kg	Critically endangered

Deep shadows fill the forest floor as towering, vine-snared trunks rush to the sky to steal the day's light before it can filter to the understory of thick green foliage. Crystaline waters pool around rocks, sprayed with splays of ferns and cushions of moss. The chatter and chirps of the forest fauna thickens in the evening air.

Suddenly an apparition at the water's edge, assembling itself from thin air and wearing the shadows themselves, Indonesia's last tiger pads its powerful way into the cool and clear pool, all-seeing eyes of verdant glass focused on the fish shimmering beneath its feet.

The critically endangered Sumatran tiger is the last surviving tiger species in Indonesia. It is the smallest of the tigers, but still a heft of powerful muscle that demands reverence from those who live alongside it. Its stripes are more closely banded together for added camouflage, rippling across its burnt-orange coat, an adaptation that helps them blend perfectly into dense thickets of the Indonesian forests. Each tiger's stripes are unique, like a zebra's, helping identify individuals.

Their powerful paws are webbed, for this feline loves a swim. They will even hunt fish, alongside things like wild boar and deer that they stalk silently from their station in the shadows, before ambushing them with a burst of rapid speed. These beasts can reach 40mph when flat out and deliver a mighty blow to stun unsuspecting prey. They sleep up to 20 hours a day, using the darkness of a jungle night to further conceal them when hunting.

Some local communities, such as those living in the village of Sekalak, have myths that tell the story of a man who shapeshifted into a tiger and became the guardian of the village.

Tragically, a mere 4,600 remain in the wild. They are under threat from deforestation as humans clear forests to make way for things like mines. They are also poached illegally, and their skins and other body parts sold on the black market.

Red Panda

Ailurus fulgens

Nigalya ponya • Habre • Punde Kundo • Firefox

Nocturnal and Crepuscular

Location	Average Lifespan	Average Size	Average Weight	IUCN Red List Status
Eastern Himalayas and Southwestern China	7 years	55–63cm (excluding tail)	3–6kg	Endangered

In the mountainous forests of the Himalayas, a copper creature tucks a thick and fluffy tail around its body, settling in to sleep peacefully, high above the leaf-littered floor. The red panda leads a tranquil life, emerging after sunset to search for bamboo and using its paws to grasp fruit and other sweet treats. They spend an awful lot of their time snoozing. They can be found by day snuggled into the nooks of tree branches, their red fur glowing in the warmth of the sun, black legs dangling from the boughs, like very furry mosses. They look like foxes that one day decided they would quite like to live in the trees, so that's what they did.

They have bear-like paws, with semi-retractable claws used for clambering through the canopy. However, they are not a bear, or even related to a panda! They even eat bamboo, and were named pandas before the more famously known black and white great panda, so which is the fake panda, really? Their closest relations are skunks and weasels, but they make up their own unique family, Ailuridae. They have very flexible ankle joints, meaning they can climb headfirst down tree trunks! Opposable thumbs give them the ability to grasp at bamboo shoots, potentially leading to their Nepali name, *nigalya ponya*, which means 'bamboo foot'. Their gorgeously ginger fur keeps them warm in cold temperatures, but climate change means they are being pushed to higher elevations and losing habitat as a result of a warming planet.

In Himalayan folklore, many stories depict these elusive and sanguine spirits as magical beings capable of bringing good fortune. In Bhutan, they are linked to a creature of myth called Balu. Balu is found often in stories in which they give guidance to travellers, protecting them as they traverse the mountains, giving the red panda a reverence as a token of protection to those lucky enough to encounter this gentle and placid panda.

Wolverine

Gulo gulo

*Carcajou • Quickhatch • Skunk Bear • Fjellfross
• Rozsomák • Tinis*

Nocturnal

Location	Average Lifespan	Average Size	Average Weight	IUCN Red List Status
Canada, China, Estonia, Finland, Mongolia, Norway, Russia and United States	7 years	45cm long	10-18kg	Least concern

What looks like a bear but smells like a skunk? A wolverine! These lone rangers prefer to seek out a solitary existence, in remote locations around the Arctic Circle. They are survival specialists, tenacious and tough with a reputation as intrepid travellers that can cover huge ranges of hundreds of miles. They may look like a bear, and are certainly packing some ferocious fangs, but they are actually the largest of the Mustelidae family, related to badgers, weasels and stoats. They have shaggy coats of brown fur, which are hydrophobic, repelling water and frost to keep this wilderness expert warm and dry, even in a blizzard.

They are quick-witted and known to think quickly on their extremely sharp toes. Like the X-Man himself, wolverines have retractable claws, but that's not the only amazing thing about their feet. They are wide and flat, acting like snowshoes that allow them to move swiftly and powerfully through the snowy tundra, their stocky but powerful legs packed with muscles that allow them to tear into frozen carcasses. Not only that, but their claws are equipped with crampons, ensuring they can carve their way through icy ground. They are adapted to life in snow and ice, meaning this species are particularly susceptible to a warming world and will struggle to survive when we lose the ice and snow.

They are nocturnal, hunting and travelling through the darkness using their sense of smell to sniff out tracks of wolves and lynx that have gone before them, hoping to find leftovers to munch on. Their scientific name translates to 'glutton', given that these survivalists leave no scraps, munching through bones and even teeth to get the most out of everything they find. Mainly they are looking for carrion, but these wandering warriors don't back down from a fight and have been seen taking on grizzly bears and winning!

Boat-billed Heron

Cochlearius cochlearius

Pico-cuchara • Boatbill • Chocuaco • Arapapá

Nocturnal

Location	Average Lifespan	Average Size	Average Weight	IUCN Red List Status
Central and South America	8 years	45–50cm long	680–770g	Least concern

By day this cryptic character holds court, hunkered down in the branches of mangrove swamps or at the edges of lakes with its feathered fellows. A closer look reveals a buff-coloured chest, leading to two bulbous and intensely black eyes that balance either side of an outrageously large black bill. A robe of grey feathers flows down his flanks. Atop of his fine-looking head, a jester's cap sits flat until raucous laughter erupts from that brilliant bill, head thrown back sending his obsidian cap feathers skyward, shaking his cap into a crown that stands proud.

Boat-billed heron are a fascinating species, by day they can be seen roosting and preening, sometimes in large groups of fifty or so individuals. But come sundown, they prefer to work alone. They leave the roost to wade silently at the shallow edges of their watery homes, standing still for an age, using their huge eyes to peer through the inky water to hunt for their prey. Once spotted, they move swiftly, using their huge bill to scoop the unsuspecting creature that unwittingly swam by.

These birds are incredibly sensitive to light and have been observed refusing to eat their prey if any light source at all was present, including a bright moon. This means this species is likely to be highly susceptible to changes in the natural night-time environment due to the growing problem with light pollution. As light pollution trickles its way into all the darkest parts of the world, natural night-time is going to cease to exist. This is a big problem for species like our boat-billed heron, who depend on natural darkness for their survival.

Kinkajou

Potos flavus

Honey Bear • Night Ape • Night Walker

Nocturnal

Location	Average Lifespan	Average Size	Average Weight	IUCN Red List Status
Central and South America	20 years	65-75cm long (excluding tail)	4.6kg	Least concern

Strutting its stuff along the tropical treetops is this silky and beautiful beast. With fur the colour of golden honey, and a feline-esque slinky stride, the canopy may as well be a catwalk to this glamourous creature. There is no denying, they have the cute factor, with a face like a bear and the body of a lemur. Is it a bear? Is it a monkey? It is neither. They are uniquely themselves, being the only species of their genus, *Potos*, with the closest relation being raccoons. They have long, prehensile tails that they can use to help climb and steady themselves as they stride along tree trunks and branches, aided by their amazing feet, which have the ability to turn completely backwards! Dexterity is crucial when you live and work at height.

Their Latin name translates to 'golden drinker', hinting at this midnight-loving mammal's eating habits. It likes to feast on the flowers of the forest, utilising their long, skinny tongues that they expertly manoeuvre to get to the nectar. They have a reputation for having a bit of a sweet tooth, loving figs and other fruits to fuel their nocturnal escapades. They are also partial to honey and are known to go on night-time raids of hives, their dense fur giving them some protection from the angry stingers of the bees they steal from.

These creatures are naturally curious and inquisitive, living their noisy lives chattering high above human heads. They are not currently deemed 'at risk' but like all animals that humans deem 'cute'. Their numbers in the wild are now decreasing as many are captured and sold on illegally as pets, with these animals being shipped out to homes around the world as humans bid to outdo each other for the most exotic or cutest pets to plaster over social media. Wildlife should remain in the wild, where it belongs, where it can enjoy its nocturnal wanderings in peace in its rightful home.

Rondo Dwarf Galago

Galagoides rondoensis

Bush Baby • Rondo Bush Baby • Nagapies

Nocturnal

Location	Average Lifespan	Average Size	Average Weight	IUCN Red List Status
Tanzania	10 years	12-13.5cm long	60g	Endangered

In the evergreen forests of Tanzania, while the Sun is overhead and burning brightly, the canopy of leaves offers cooling shelter and shade to a creature that prefers to clamber around the treetops under cover of darkness than in the harsh light of day, so instead she snoozes soundly, in her nest away from harm.

The Rondo dwarf galago is a palm-sized primate, and the smallest of all the galago (or bush baby) species, weighing the same as a tennis ball. They are incredibly agile and can move quickly, using their long fingers and toes to curl around tree branches, leaping effortlessly from one perch to the next. They have perfected the night-time adaptations, with two huge eyes that glow amber, able to detect the smallest of movements on even the darkest of nights. A snub, pointy nose is painted with a white strip, leading to the top of its grey head where two bat-like ears sit, primed and ready to act like radar. They can move each ear independently, allowing them to track insects at night, and have such good hearing that when it's time to sleep they roll their ears down, to lessen the daytime din of their forest neighbours. Being nocturnal, they are hard to spot in the wild, you might be lucky enough to catch their eyes reflecting your torchlight back from the depths of the bush, or catch sight of its bottlebrush tail as it springs gracefully away into the night, but you will more likely hear their wailing cries in the night. Galagos are called *nagapies* in Afrikaans, meaning 'little night monkeys'.

Sadly, these darkness devotees hold an unfortunate title, as one of world's most endangered primates. Their habitat is at risk due to logging and other factors that have seen their leafy lairs torn to the ground. Work is underway to create a network of reserves in Tanzania, to give these miniature monkeys the protection they deserve.

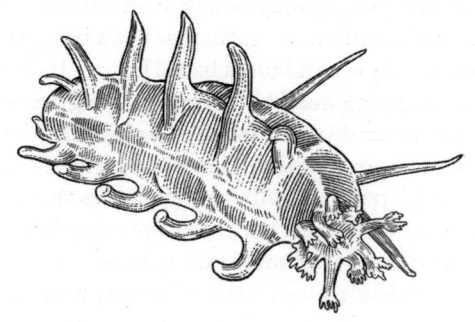

Sea Pig

Scotoplanes globosa

Nocturnal

Location	Average Lifespan	Average Size	Average Weight	IUCN Red List Status
Arctic, Atlantic, Pacific and Indian Oceans	Unknown	2–15cm	Unknown	Not listed

The world's oceans cover over 70 per cent of the surface of our planet, to view the Earth from space is to see that it isn't really 'earth' at all, it is oceans, seas, rivers and lakes. We are water, it is what made this planet habitable. Those oceanic depths hold 97 per cent of the Earth's water, gathered in gargantuan goblets that descend to the deepest points of the world's surface.

Craggy cliffs, marine mountains, and cavernous canyons conceal violent volcanoes, abyssal plains, and a myriad of complex ecosystems and life forms.

The sea has long held the human imagination, ensnared in a net that has been cast out and captured the minds of our ancestors for eons, inspiring poetry and song. But not even the human imagination can conjure the weird and wonderful life forms that have inhibited the hidden world of the sea floor.

Enter stage left; the sea pig. Resembling a pink tubular bag of jelly with legs, these creatures are a species of sea cucumber, which are not actually cucumbers, but 'echinoderms', a group containing saltwater surprises like starfish and urchins. Living in the deepest parts of the ocean, it spends its life in total darkness, using hydraulics to propel its way through life, sniffing out fresh dead things that have fallen from above. They are particularly partial to things like 'whale fall'. When a whale dies, it will fall to the bottom of the ocean, where it becomes a buffet for all the opportunistic floor feeders. The carcass contains a huge amount of nutrients, which sees sea pigs arrive in their hundreds to get their fill. They are hugely abundant, and in some parts of the ocean these gelatinous grafters make up 95 per cent of the total weight of animals living on the deep-sea floor.

Sea pigs can also play host to parasitic crabs, which attach themselves to a sea pig and ingest its organs. Just in case it couldn't get any weirder, living this far down requires some unusual modifications to ensure vital bodily functions can take place. Like breathing through its anus.

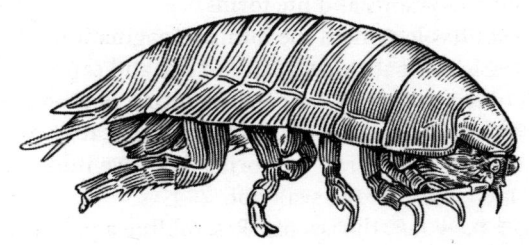

Giant Isopod

Bathynomus giganteus

Nocturnal

Location	Average Lifespan	Average Size	Average Weight	IUCN Red List Status
West Atlantic and Indo-West Pacific	Decades but unknown	40-50cm	3-3.5kg	Not listed

If you were to go to your garden and take a peek under a plant pot or a rock, chances are you would see the frantic scuttling of woodlouse as you exposed a hidden, dark and damp world to daylight. Now, imagine you're enjoying a leisurely journey along the dark of the seabed when you see a giant woodlouse scuttling towards you!

That essentially is what giant isopods look like, and are indeed, related to, albeit a lot heftier.

With fourteen legs and a tough, lilac-grey 'shell', these isopods trundle around the seabed, trekking into the deep and treacherous trenches like armoured marine tanks. Some species of isopod can even inhabit the incredibly hostile hadal zone, named after Hades, the Greek god of the underworld. They look positively alien, with large, triangular eyes built to reflect the miniscule amount of light that is available at the bottom of the big blue.

The environment this darkness-dweller calls home is dark and incredibly cold, with temperatures of around 4°C and the pressure is literally, crushing. They are important to our ocean's ecosystem, serving the seas as scavengers of the sea floor, clearing up all the detritus (known as detritivores) of dead things that succumb to the depths, releasing valuable nutrients that go back to supporting other life forms in our marvellous marinas. They have adapted over 300 million years to be survival experts, as waiting for your food to fall from the sky can require lots of patience. One isopod was recorded as leaving 5 years between meals!

Our oceans contain the deepest and darkest places that exist on the planet, that conceal mysterious life forms that exist totally unbeknown to humankind, as we have only explored around 5 per cent of the ocean. Who knows what else is down there?

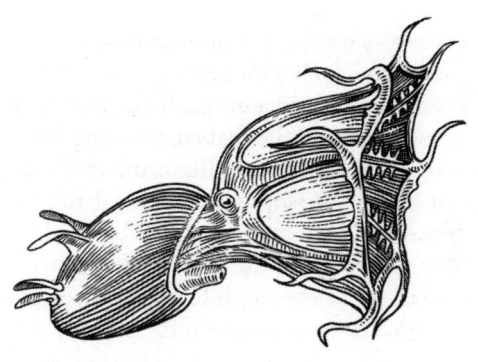

Vampire Squid

Vampyroteuthis infernalis

Nocturnal

Location	Average Lifespan	Average Size	Average Weight	IUCN Red List Status
Global	8 years	30cm long	0.5–1kg	Not listed

As things that live in the deep dark go, and carrying a moniker like the 'vampire squid from hell' as the Latin name translates, you would think this creature is some sort of infernal and wretched being, with huge, sharpened jaws ready to stalk out of your nightmares to drag you down into the unknown and suffocating depths of the ocean.

If you're into your horror, prepare to be disappointed. This squid is nothing but a damp squib when it comes to terror. Firstly, it's not even a squid. It doesn't even eat live prey, and is the only cephalopod known not to. In fact, I think we need to have words with whoever named this poor thing in the first place as, in reality, it is more fabulous than it is frightening.

They are an ancient being, evolving from an octopus ancestor some 165 million years ago. Living under the waves at depths ranging between 600 to 1,200 metres, this showbiz squid floats around the ocean's midnight zone, an ecological area where sunlight doesn't dare to shine; so instead the creatures that reside here can sparkle. What this dainty Dracula lacks in bloodlust, it makes up for with theatrics.

A deep red in colour with two enormous, show-stopping eyes (the largest proportional eyes in the animal kingdom), with two fins either side of its head that give it a curious resemblance to Dumbo. Delicately dancing through the deep, it lives on a diet of marine snow. It has a luxuriously plush cape, like webbing, that connects its eight arms. When at risk of predation, like a magician performing a trick, it can swoop its cape over its body to be shrouded. Its body is covered in luminous photophores. These are light-emitting organs which can dazzle a would-be predator. Failing that, a final flourish can see this showman eject a plume of bioluminescent plankton, helping the vampire squid complete its disappearing act.

Gilbert's Potoroo

Potorous gilbertii

Ngilkat

Nocturnal

Location	Average Lifespan	Average Size	Average Weight	IUCN Red List Status
Western Australia	8 years	55cm long (including tail)	1–1.2kg	Critically endangered

A small being scuttles around the dense scrubland in the dead of night, a pointed, inquisitive face is barely glimpsed as this rodent-looking marsupial rummages around for its favourite fungi. At first glance you may assume the creature in the bush is a rat, but look closer and you will see a chubbier face, and dense silky russet fur from which springs hind legs that look suspiciously like a kangaroos, she even has a pouch too! Half-hopping, half-scurrying, the Gilbert's potoroo is sometimes given the nickname of 'kangaroo-rat', but if you're lucky enough to glimpse one, you will have seen Australia's most endangered marsupial, and in fact, the rarest marsupial in the world.

It seems that fate has never been on the side of this poor little soul. We thought it was long lost, to the overzealous colonial collectors from Australia's past. It was a complete surprise when it was rediscovered in 1994 after being presumed extinct for 120 years. The first 'specimen' was collected in 1840 by taxidermist John Gilbert, who travelled from England to study and collect animals as part of the colonisation of Australia by Western Europeans, and from there on in, times got a lot trickier for this unassuming being.

Their population numbers have plummeted, thanks to things like introduced species that colonisers brought with them to Australia such as foxes and cats, things that a potoroo had never seen and never had to learn to live with before, so they have been predated and practically wiped from existence as a result.

Then in 2015, a large fire destroyed 90 per cent of their only remaining natural habitat and killed fifteen of the twenty potoroos living in the reserve. Luckily, alongside the very small population in Two People's Bay, conservationists had already introduced potoroos to a few outlying islands where trials are underway to try and boost the population. Without this work the species would have been gone for good, but now the population is around one hundred individuals with people hard at work to keep this species on the planet for good.

Axolotl

Ambystoma mexicanum

Water Monster

Nocturnal

Location	Average Lifespan	Average Size	Average Weight	IUCN Red List Status
Mexico	5-6 years	40-45cm	230g	Critically endangered

There is no other creature that looks more like it should belong to myth than reality than the charming axolotl. They look like small, cartoon water dragons, with round comical faces that appear to be smiling and feathery fronds branching from their necks like mini corals. They are a type of fully aquatic salamander, totally unique of other salamander species due to their ability to live their entire lives underwater. Despite having fully formed lungs, their feathery fronds are actually gills that enable them to lead their aquatic lives. They come in an array of colours, ranging from midnight black to albino, with some axolotl that almost appear translucent. If you think they look familiar, they were the inspiration behind the dragon Toothless in *How to Train Your Dragon*. It isn't hard to see the likeness.

They are exceptionally rare, surviving only in a very specific habitat. They can be found in the ancient Lake Xochimilco and its canals in Mexico City, their last remaining home on Earth. In 1984 the Mexican government declared their last bastion a protected biological reserve, and UNESCO deemed it a World Heritage Site just 3 years later.

Even more fantastical sounding is their ability as the only vertebrate to regenerate limbs, being able to regenerate an entire leg in just a few weeks. Not only limbs, but internal organs and even parts of its brain. This has led to widespread interest by medical researchers to try and discover cures for human ailments, such as types of cancer.

The main threats to its survival are human influences like water pollution, destruction of habitat, competition from non-native species, and the illegal wildlife trade. Axolotl are completely nocturnal and have eyes that are very sensitive to light sources. They are active during hours of darkness and lighting can stop them being able to function naturally. During the day they like to hide away from predators like heron.

Legend has it that the Aztecs named axolotl after Xolotl, their deity of lightning and fire. The story goes that to avoid becoming a sacrifice so that the Sun and Moon could move in the sky, Xolotl used the power of shapeshifting to become an axolotl.

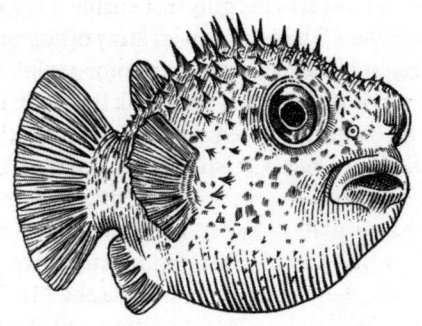

Spot-fin Porcupinefish

Diodon hystrix

Ajargo • Uruhakau • Tagotongan • Sokisoki
• 'O'opu-kawa

Nocturnal

Location	Average Lifespan	Average Size	Average Weight	IUCN Red List Status
Eastern Pacific, Western Atlantic and Western Indian Ocean	10 years	40cm long	2.8kg	Least concern

Swimming along rocky reefs in colourful and tropical waters around the world, you may come across this iconic fish species. When calm and tranquil, their body resembles the shape of a baseball bat, with a wide, large head that tapers into a much narrower tail. It has large and dark eyes, set either side of its comically large-lipped mouth, which seems to be downturned into a grumpy sulk. Its skin is a light sandy grey with a white belly covered in small black spots that help camouflage it into the rocky reefs it calls home. But of course, it is most famous for the spines that cover its entire body and its incredible 'party trick'.

When threatened, this fish has developed the ability to make itself look as intimidating as possible. All along its body are specially adapted scales that can be weaponised in defence. If threatened, the fish will inflate itself, by taking in water to rapidly make itself twice its normal size and causing its fearsome-looking spikes to stand on end. They are even missing some ribs and have no pelvis, so they can do this defence mechanism without the fear of breaking any bones. This technique should deter any predator, but if one was trying its luck, the porcupinefish makes itself so rigid that it would be incredibly difficult to eat. They also produce a toxin which makes their internal organs extremely poisonous. Despite this, they are still eaten and considered a delicacy in places like Japan, where it is known as *fugu* – chefs must train for years and pass multiple exams before being able to serve the fish as a dish.

They are a solitary and nocturnal species, preferring to wander the dark ocean evenings alone. They have a single 'tooth', which is really lots of teeth that have fused together into a single unit. This means this fish has a very strong mouth, more like a beak, that is able to break through the hard exterior of snails and crabs.

Aardvark

Orycteropus afer

Earth Pig • Thakadu • Sambane • Ihodi • Xikhwari

Nocturnal

Location	Average Lifespan	Average Size	Average Weight	IUCN Red List Status
Sub-Saharan Africa	18 years	1-1.3m long (excluding tail)	60-80kg	Least concern

Aardvarks are nocturnal, burrowing mammals that live in a wide variety of habitats including savannahs and rainforests throughout Africa. They look like a jumble of lots of animals put together: long ears like a rabbit that stand on top of its head, a lengthy tail similar to a kangaroo, and a sensitive snout that wouldn't be amiss on Babe the pig. It's the swinelike nose that gives it its common name, 'aardvark' being Afrikaans for 'earth pig'.

They are a very secretive creature who spends all day sleeping in their burrows, choosing the cooler air of a starry night to go foraging for ants and termites. They can travel far across the darkness, covering as much as 30 kilometres a night. Their eyesight may be poor, but they have a powerful nose and large ears which help them home in on termite nests. Termite nests can be huge, very solid structures up to 10 metres in height, they stay intact for long periods, sometimes centuries! But the aardvark is prepared. It is armed with long, spoon-shaped claws that can easily break into the termite mound, creating a hole through which the Aardvark will shove its snout, unleashing its sticky foot-long tongue to collect up to 50,000 termites in a single evening.

Despite being similar in appearance and feeding habits, aardvark are not related to the anteater, they're actually more closely related to the elephant. They are so weird and wonderful, that they are one of the species that the science community considers 'living fossils'. Their genetic make-up has remained unchanged for over 50 million years, giving us insight into species that are now sadly long gone.

Along with many other burrowing species, their burrows are important ecological features, and can be used by other species when they need protection from predators or wildfires. Warthog, porcupine, and even snakes have all been recorded as using their underground lairs. They play an important part throughout folklore across many African countries. It is thought that the Egyptian god Set, who rules over desert, storms, and chaos, is often depicted wearing the head of an aardvark.

Plains-wanderer

Pedionomus torquatus

Nocturnal and Diurnal

Location	Average Lifespan	Average Size	Average Weight	IUCN Red List Status
New South Wales and Western Queensland, Australia	Unknown	15cm long	98g	Critically endangered

Australia is a continent that is home to a myriad of unique and, sometimes, quite baffling species of wildlife. There are perhaps none more curious than the peculiar plains-wanderer. It makes its home in the grasslands of Australia, one of the continent's most threatened habitats thanks to pressures from agriculture and wildfires. At first glance, you probably won't be captivated by the bird's plain plumage. Not too dissimilar to that of a quail or female pheasant, they're a brown and buff bird, with a speckled chest akin to a mistle thrush. A beady yellow eye will sometimes be seen as it cranes on tiptoe, trying to suss out whether to sit tight or flee from a threat. A slender yellow beak and two long legs in a matching shade, are nothing that's going to get a purveyor of exotic and colourful birds going. But this bird is anything but dowdy, and it is in fact one of Australia's rarest birds.

There is potentially only just over 250 of these birds left in the wild. A timid and, rightfully, anxious avian, they are not good fliers. If a threat appears its only real option is to run away. Living out in open grasslands it can be difficult to find cover to hide, which makes it easy prey for introduced predators such as foxes, domestic dogs and cats.

It nests on the ground, creating a shallow scrape in the dusty earth to deposit its eggs. Just because it likes to be different, it's the males who incubate the eggs on the nest. More peculiar still, there is nothing else like it in Australia. It is one of a kind, belonging to an ancient family of birds Pedionomidae and does its its best to look like any other grassland-dwelling species, but it is actually a type of wader, or shorebird. Little is known of its nocturnal habits, but it appears to be predominantly active in the night, foraging for food under the relative safety of darkness.

Red-eyed Tree Frog

Agalychnis callidryas

Rana de árbol de ojos rojos • *Grenouille aux Yeux Rouges*

Nocturnal

Location	Average Lifespan	Average Size	Average Weight	IUCN Red List Status
Central and South America	5 years	5-6.5cm long	12-15g	Least concern

A cursory glance at the red-eyed tree frog would instantly explain the name. This petite amphibian makes up for its size with two huge red eyes that sit, bulbously protruding from either side of its bright, lime-green head. When asleep, its special 'eyelid' is actually a thin membrane, allowing it to camouflage itself while still allowing enough light in so that if a predator approaches, the shocking flash of his peculiarly red eyes will startle them, buying these frogs precious seconds to escape to safety.

They say you have to kiss a lot of frogs to meet your prince, but this frog is practically a carnival king, robed in exotic colours of sapphire, citrine and amber that help blend into his tropical surroundings. At the end of his long, smooth limbs built for climbing trees rather than swimming, sit large, webbed feet, in a spectacularly eye-catching orange. At the end of each of his four toes, chalice-shaped footpads that create a strong suction between frog and leaf, so strong they can actually walk upside down – a handy feature when needing to lay your eggs on the underside of a leaf! A very clever behaviour, these leaves overhang ponds as well as other water sources, and as the young tadpoles need water when they hatch, they will simply fall into the water directly below them.

This species is completely nocturnal, spending its days curled up safely on the underside of a leaf. By night it's on the hunt for insects such as grasshoppers and sometimes even smaller frogs, using its long sticky tongue to capture prey. Its colouration is very bright, but unlike other brightly coloured amphibians it isn't toxic. This frog is faking! The luminosity of these colours helps to fend off would-be predators, most of whom are nocturnal like owls and tarantulas. The bright colours are more perceptible in the darkness, so any hunters may be tricked into thinking their tasty treat is actually toxic. There's no call for style over substance here!

Wondiwoi Tree-kangaroo

Dendrolagus mayri

Nocturnal and Diurnal

Location	Average Lifespan	Average Size	Average Weight	IUCN Red List Status
Wondiwoi Peninsula, Papua New Guinea	Unknown	50cm long	9.25kg	Critically endangered

Papua New Guinea is rich in diverse and fascinating flora and fauna, housing more than 5 per cent of the world's total biodiversity in less than 1 per cent of the world's land mass. Much of the life found in this beautiful country is incredibly rare, with many species found nowhere else on the planet. One of those twinkling treasures was long considered extinct, the Wondiwoi tree-kangaroo.

High in the cloud forests, the plant life shimmers in hues of emerald, jade and garnet, while fluffy white whisps shroud the canopy, misting everything underneath with lifegiving water. Up at an altitude over twice the height of the UK's largest peak, Ben Nevis, lie forests so dense that they are impenetrable. It is here that some of the Earth's most precious species can be found.

Very little is known about the behaviours and habits of the Wondiwoi tree-kangaroo. The only specimen known to science was shot in 1928 and sent to London's Natural History Museum, this single specimen was all we knew about this creature. It was over 90 years before it was seen again, a fleeting photograph taken by an intrepid explorer. Imagine the lovechild of a kangaroo and a koala, and you will come close to how this animal looks. For decades we previously only had drawings to tell us what this arboreal marsupial looked like.

Part of the Macropod family, which includes wallabies, quokka and kangaroos, these animals would have all once, way back in their evolutionary history, lived in the treetops until they came to love a life on the ground. However, tree-kangaroos represent a quirk that sometimes happens in evolution, after coming to live on the ground they decided they would rather return to the canopy, and there they have stayed. Nocturnal in nature, they live out their peaceful lives away from the world's gaze, leaping from tree to tree, using their long and muscular tails for balance. Now their home is under threat, with greedy eyes looking to their habitat to open a gold mine, which would destroy the natural treasures that live in these forests. With just an estimated fifty individuals left, there can be no chances taken to ensure these curiosities' survival.

Greater Capybara

Hydrochoerus hydrochaeris

Ka'apiûara • Kapii'gwara

Nocturnal and Diurnal

Location	Average Lifespan	Average Size	Average Weight	IUCN Red List Status
Central and South America	10 years	1.3m long	40–66kg	Least concern

Essentially a barrel with legs, the greater capybara is the world's largest rodent. They essentially look like a giant, aquatic guinea pig or a beaver without its tail. They are bigger than you may first imagine, roughly reaching the size of a border collie and weightier than some humans. Covered in luxurious auburn locks, they wouldn't look out of place in an advertisement for shampoo, which judging by the internet's obsession with 'capybaras in spas', wouldn't be too incredulous to imagine. They are semi-aquatic, spending the hot and humid days of the Amazon wallowing in shallow waters to keep themselves cool, preferring to dine after the Sun has set, giving the added advantage of some protection from keen-eyed predators who wouldn't mind a tasty capybara being on the menu.

Their feet are slightly webbed, giving them an added advantage in their watery surroundings. To escape predation from the likes of jaguars and eagles, they can remain submerged under the surface of the water for up to 5 minutes at a time, leaving enough time for the predator to give up and move on to an easier thing to hunt. They like to lead a pretty chilled lifestyle and can often snooze while floating, keeping their snout above water like a fuzzy snorkel and bringing a new meaning to a 'lazy river'.

They're a highly social species and like to let you know that. They like to keep in touch through a variety of vocalisations that span purring and barking to teeth chattering. Their name originates from the indigenous language of the Tupi people, with *ka'pii* meaning 'grass' and *gwara* meaning 'eater', perfectly describing the eating habits of the vegetarian king of the rodents. They are an incredibly endearing creature, known for being gentle and accepting of other wildlife species, including crocodiles. For reasons unknown, crocodiles won't eat a capybara and they can often be seen hanging out together, including unbelievable footage of capybara riding on the backs of crocs up a river. Some suggest capybara are good at alerting the crocs to potential predators as a reason for this unlikely friendship, but perhaps the capybara is just the true social butterfly of the animal world.

Southern White Rhinoceros

Ceratotherium simum simum

Chipembere • Tshukudu • Witrenostre • Ubejena

Nocturnal and Diurnal

Location	Average Lifespan	Average Size	Average Weight	IUCN Red List Status
Sub-Saharan Africa, South Africa, Kenya, Uganda, Zambia, Mozambique	35-40 years	3-4.8m long	1,800-2,700kg	Near threatened

A reticent and reserved giant ambles its way across the African plains, stocky shoulders swaying, pewter-plated in thick and toughened skin. Two teardrop-shaped ears swat away insects that dare to interrupt this humble Hercules as its broad head browses the short grasses of the savannah. A furrowed brow slopes into two horns, the smaller sits between two eyes that placidly peer across the plains. Erupting from the top of the mountainous muzzle, a second much larger horn leads the charge of this vegetarian mammoth mammal. Despite the stout appearance of a circus strongman, the white rhino can be fast, reaching speeds of 31mph if threatened.

The largest of the rhino species, they are also the most social, living in small 'crashes' of up to fourteen rhinos. Misleadingly named, these rhinos are actually grey – this confusion thought to have occurred when English speakers mistook the Afrikaans term *wyd* meaning 'wide', referring to the mega-sized muzzle.

There are African beliefs that this majestic creature is a symbol of Mother Earth herself, appearing in folklore as the guardian of the land and its people. The horn is symbolic of authority, courage and protection. When Europeans first encountered this beast, they couldn't believe their eyes, with only word of mouth and some (very) liberal artistic impressions to go off, the rhino became legendary as an armoured, semi-mythical beast. The last known rhino in Europe was at the time of Ancient Rome, so the stories of this survived and at some point, rhinos and unicorns overlapped in mythical lore.

Maybe this began due to the belief that rhino horn holds magical, healing powers (an idea that survives to the this day), leading to the tragic demise of this beautiful beast. Their horns are not ivory, but are touted as a luxury health tonic, something that is a bigger myth than the unicorn itself, leading to illegal trade in rhino horn. The northern white rhino now has only two females in existence. They will go extinct. So abhorrent are humans that the hopes of this species lay at the feet of a four-year-old rhino, who was brutally murdered after poachers broke into a zoo in Paris, removing his horn and the hopes of the recovery of this species.

African Civet

Civettictis civetta

Siwetkat • Tshipalore • Afrikaanse civet

Nocturnal

Location	Average Lifespan	Average Size	Average Weight	IUCN Red List Status
Sub-Saharan Africa	15-20 years	70-84cm long (excluding tail)	10-20kg	Least concern

What is a civet? Slinking along the forest floors of Africa at night, the bright beam of your torch may illuminate a familiar shape and movement. Your mind will immediately be cast back to the neighbourhood tabby, perhaps a very large version, a striped and spotted coat that overlays black on to grey, its ringed tail sweeping low through the grasses, blending this foraging faux feline into its surroundings perfectly. Then something catches its attention, and it turns to face you, now caught head on in your artificial light its face is finally revealed, but you are presented with a mystery, this cat you see has got the head of a raccoon. Complete with shiny nose and bandit face mask, it quickly saunters back to the safety of the shadows. This isn't a cat at all. Nor is it a raccoon. The civet is a curiosity of its own design, belonging to the family 'Viverridae'. They are more closely related to weasels than anything feline.

Perhaps going through a punk phase, this unique individual sports a large and bristly Mohawk that clings to the length of its spine, known formally as a 'dorsal crest'. This is used when threatened to stand on end, making the civet look as big and scary as possible to fend off anyone looking for a fight. In keeping with their style choices, they apply the same anarchistic approach to their meals. The civet can eat creatures that are normally far too toxic for other mammals to nibble on, including a poisonous millipede which contains cyanide!

Humans have found a lucrative use for civets, using their secretions from anal glands in the perfume industry to create 'civetone', a strong musky odour that is used in fragrances, leading to a cruel trade where civets are kept in tiny cages, where they will have their glands scraped. Many perfume manufacturers now choose the more ethical synthetic version, but some still choose the inhumane way, so always check your perfume doesn't contain animal-derived civetone.

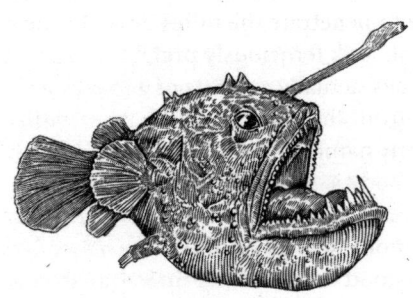

Deep-sea Anglerfish

Lophius

Sea Devil • Devilfish • Oof • Cythraul y Môr • Maisgoom • An Láimhíneach • Diabo Marinho

Nocturnal and Diurnal

Location	Average Lifespan	Average Size	Average Weight	IUCN Red List Status
Global	25-30 years	up to 1.2m	up to 50kg	Least concern to near threatened

Talk about a fatal attraction. This alluring deep-sea dwelling fish comes equipped with its own actual fishing lure. Living down to the depths of the sea floor, this is a creature that is made for darkness, living its life out in the 'Midnight Zone' of the ocean. Our oceans contain some of the deepest dark on the planet, the light of the sun unable to penetrate the miles down to the seabed.

These fish look ferociously prehistoric. The order 'Lophiiformes' actually consists of 200 species of anglerfish, from the 'Dreamer' with its ultra-black skin to the adeptly named 'Black Sea Devil'. A huge blob of a dark grey body is armed with a giant upside-down crescent mouth, lined with a terrifying set of teeth. It is the face that only a mother could love, like a bloated toad that's been stood on. Its dorsal fin is translucent, with silver spines, moving in a serpentine fashion down its spine to a final flourish of a feathery tail.

Sexual dimorphism is the physical difference in appearance between males and females in the species, and in the majority of anglerfish the males are tremendously, tiny. Miniscule. In comparison, the females look like a She Hulk. Proving that size really doesn't matter in the face of love, the Kroyer's anglerfish female can reach up to 1.2 metres in length, compared to the males' mere 16 centimetres. Their love is all encompassing, quite literally, with the males being parasitic, and upon latching to his giant love he will slowly eternally fuse himself to her, losing his brain, eyes and pretty much everything else bar his reproductive organs in the process, bringing a new meaning to 'til' death do we part'.

When you inhabit some of the darkest places on the planet, light is something of a spectacle. This is something the anglerfish have known for millions of years and have adapted the power of light, using it to their advantage. The females have their own lures, each one unique to the individual, with some species having multiple. It's the irresistible light of this lure that draws a multitude of fish and crustacean species to their demise, enchanted by the beauty, they are oblivious to the ugly death that lurks just behind.

Eurasian Beaver

Castor fiber

*Afanc • Dobhran Losleathan • Béabhar • Bäver• Castor • бобр •
Bóbr • 海狸*

Nocturnal and Crepuscular

Location	Average Lifespan	Size	Weight	IUCN Red List Status
Scotland, England, widespread across Europe into Mongolia, China and Russia	8 years (25 max)	0.8-1m long (excluding tail)	up to 30kg	Endangered (UK), Least concern (global)

No rodent is more charismatic than the loveable beaver,
second only in size to its larger cousin, the capybara.

Known for its natural instinct to engineer dams, they have featured in a wide range of literature and folklore that spans the ages, although perhaps their most notable appearance is that as 'Mr and Mrs Beaver' in *The Lion, The Witch and The Wardrobe*. As famous as they are, they have been widely misunderstood, with common myths about their behaviour seeping in through folklore and stories, such as anglers believing they will eat up all the fish in their chosen lake or pond, despite these critters being strictly vegetarian.

Beavers were once widespread across Europe, but they have been hunted for thousands of years for their waterproof fur and castoreum, an oily secretion from their anal sacs. This is sometimes used in food products, with food manufacturers preferring the massive umbrella term 'natural flavourings' rather than 'beaver anal sac secretions'. It's even used in make-up products and perfumes. Due to their usefulness, beavers were hunted to extinction in the UK around the late seventeenth century. They left our landscapes, and our collective memory, with the word 'beaver' being replaced by 'blog' in the Oxford Junior Dictionary in 2008.

Beavers are mainly nocturnal, preferring the hours after dusk and before dawn to go about their lives under the light of the Moon. During the day they will be snoozing in the safety of their lodges. They create dams by gnawing at riverside vegetation, coppicing trees as they go and dragging the branches to slow down the flow of the water, creating an area of still water, a pond. Behind their dams they will then create their lodges, a pile of sticks and mud with an underwater entrance, to keep them safe from predators. These ponds create vital habitats for all sorts of other wildlife, too.

Hawksbill Sea Turtle

Eretmochelys imbricata

Honu'ea • Tortuga Carey • Vonu • Kasa kingange • Taku• Crwban Mawr Gwalchbig • Honu Afii Moa • Sagr

Nocturnal and Diurnal

Location	Average Lifespan	Average Size	Average Weight	IUCN Red List Status
Atlantic, Pacific and Indian Oceans	60+ years	1m long (shell)	65kg	Critically endangered

Gliding serenely through the warm and crystalline waters, flippers outstretched like sea wings, a keen eye searches among a carnival of coral reef, a riot of colour that ebbs and flows, giving rhythm to the lungs of the ocean. A kaleidoscopic carapace shimmers in the sun beams like scintillating aquatic aurora. This coy creature silently surveys until it finds a favourite snack, a leathery barrel sponge. Using its sharp, hawk-like bill, it is prised from the reef and eaten.

One of the smaller of the seven sea turtle species, the hawksbill sea turtle is also one of the world's most endangered reptiles. Living among the coral reefs, it is vital to maintain the health of these incredibly intricate and important ocean habitats. A single turtle can eat an estimated 450 kilograms of sea sponge a year. This is a rather unusual culinary choice; very few animals on Earth are able to penetrate the sharp defences of sponges which render them toxic and indigestible to most other species, it is the only reptile with this ability.

A solitary species, they cover territories spanning thousands of miles of open ocean, only coming together to mate. They must return to land to lay their eggs in nests on beaches under the protective cover of darkness. That's where the danger lies. Humans are increasingly moving to the world's coastlines, developing what was once pristine darkness into multicoloured metropolises that are among some of the most light-polluted places on the planet. Light pollution is a big problem for our sea turtles, who need darkness to protect them from predators whilst in the act of egg laying. It is also crucial for newborn sea turtles, hatching alone up to 2 months later. They will hatch at night, where the celestial light of the stars and Moon should be a natural cue that instinctively the hatchlings will move towards, bringing them into the safety of their new ocean home. Light pollution is causing sea turtles to go the wrong way, drawing adults and hatchlings alike into resorts and residential areas, with many tragically dying, crushed by vehicles or dying from dehydration. You can help by adopting dark sky lighting principles at home.

Astronomer

Stella servator somnolentus

Star Gazer • Space Scientist • Astromancer • Seryddwr
• Star Lord

Nocturnal

Location	Average Lifespan	Average Size	Average Weight	IUCN Red List Status
Global	73 years	1.6–1.7m tall	60–80kg	Not listed

Astronomer, Stargazer, Astrologer. There are many
names for these Homo sapien subspecies that look to
the sky and have done for thousands of years. They are
easy to spot and tell apart from the wider population,
characterised by their skin usually having a paler hue
than their peers and large dark circles that mark under

their eyes, due to the nocturnal habit they follow, but this means they are seldom seen and harder to spot as you must wait until the hours of darkness to find them.

Some struggle to form social bonds as they can lead quite a solitary lifestyle, struggling to find others of their species who are willing to stay awake, often for unnaturally long hours for their species, as they fight the natural urge to sleep in hours of darkness in order to look at twinkling lights and things like moons. They are plagued by chronic issues such as painful necks and backs, and can quickly become addicted to caffeine and weather forecasts. They display a strong dislike of clouds and can usually be found nodding off during social gatherings or snoozing through the heat of the day in darkened burrows they call beds. Finding a mate can prove difficult, due to them often choosing to exist in remote locations, but courting displays can be seen through the gifting of shiny space rocks, space dust, and sometimes even telescopes to the terrestrial wonder that pulls their gaze from the skies.

But for all their quirks, they offer a glimpse into a world unknown to many. Inquisitive minds and sharp eyes set this species apart from others and they offer a lot to the world they inhibit. Although frequently ignored by others in their family, they can further improve the lives of everything on the planet, if only they could stay awake long enough to communicate their findings.

Useful Organisations and Institutions

The International Dark Sky Association
Home of all things Dark Sky. The International Dark Sky Association (IDA) started in 1988 and is now the recognised global authority on protecting Dark Skies, offering outreach, support and technical lighting advice. Their reach now extends to fifty-one countries around the world. If you want to start on your Dark Sky journey, here's the best place to begin.
https://www.darksky.org/

Prosiect Nos
The North Wales Dark Skies Partnership between Eryri National Park and the three Areas of Outstanding Natural Beauty – Ynys Môn, Pen Llŷn and Bryniau Clwyd a Dyffryn Dyfrdwy. Here is a one-stop shop for all things Dark Skies in North Wales, including the Welsh Dark Skies Week held each February.
https://www.discoveryinthedark.wales/project-nos

Eryri National Park
Home to Cymru's largest International Dark Sky Reserve and the base of operations for Prosiect Nos – The North Wales Dark Skies Partnership. Here you will find advice on where to stargaze, how to become Dark Sky friendly

and more about some of the stories in the stars you found in this book. Their website is fully bilingual and is a great place to find out what else is happening in one of the UK's original first three national parks.
https://snowdonia.gov.wales/

Bryniau Clwyd a Dyffryn Dyfrdwy/Clwydian Range and Dee Valley Area of Outstanding Natural Beauty
The Clwydian Range and Dee Valley AONB are part of Prosiect Nos – North Wales Dark Skies Partnership – and run events year-round with their fantastic team of dedicated rangers. You can even join them for a hot chocolate at their shepherd hut on Moel Famau before a Dark Skies walk in winter. Together, the partnership created this easy-to-use guidance on becoming Dark Sky friendly, with a range of examples, from home to barn!
https://tinyurl.com/Dark-Skies-Guidance

Ynys Môn Area of Outstanding Natural Beauty
Another partner of Prosiect Nos, Anglesey AONB hugs nearly the entire island's coastline. The AONB team are a tenacious bunch, working hard with land owners, home owners and businesses to change lighting to protect the Dark Skies that are at risk of being lost over the island.
https://www.anglesey.gov.wales/en/Residents/Countryside/Areas-of-Outstanding-Natural-Beauty-AONBs/Anglesey-Area-of-Outstanding-Natural-Beauty-AONB.aspx

Pen Llŷn Area of Outstanding Natural Beauty
Home to Europe's only Dark Sky Sanctuary, Ynys Enlli, and the final partner of Prosiect Nos, Pen Llŷn AONB is

situated under one of Cymru's Darkest Skies, reaching out towards Ireland. It's one of the darkest places you can get to without getting your feet wet! A small team works hard to keep the 'O' in AONB.
https://www.ahne-llyn-aonb.cymru/Home

Ynys Enlli Trust
Europe's one and only International Dark Sky Sanctuary and home to the fantastic Bird Observatory, Enlli is open to visitors between March and October. You'll need to book an overnight stay to ensure you get to see a magical night sky unfold in front of you, and your only neighbours will be seals and sea birds! Be warned, sometimes you can get there but not back again, if the weather changes.
https://www.bardsey.org/visit

Bardsey Bird Observatory
The Bardsey Bird Observatory has been keeping a keen eye on the sky since it opened in 1953. You can even join them for a night-time walk to visit their famous Manx shearwater.
https://www.bbfo.org.uk/

British Astronomical Association
If you're a budding stargazer but not sure where to start, head over to the British Astronomical Association website where you will find loads of advice. They deliver outreach astronomy sessions and have a monthly 'observing challenge' to get you stuck into space.
https://britastro.org/

Campaign for Dark Skies
A branch of the above, the Campaign for Dark Skies was created in 1989 and has been fighting ever since for better lighting to protect our night skies for astronomy. They have lots of great advice and information on their website on how you can be Dark Skies friendly.
https://britastro.org/section_information_/dark-skies-overview/issues/about-the-commission-for-dark-skies

Natural Resources Wales
Natural Resources Wales are responsible for looking after the environment in Wales. Jill Bullen and her team worked to create a wonderful map detailing the light pollution in Wales. Want to see how dark it is where you are or where you are going? Head here!
https://luc.maps.arcgis.com/apps/dashboards/1cd6ba8a1d7d4a62aff635cfcbaf4aec

Light Pollution Map
Sadly, not all of us live in Cymru. If you want to see how dark your area is, or in fact, anywhere on the globe, check out this light pollution map.
www.lightpollutionmap.info

Campaign for the Protection of Rural England
The CPRE do all sorts of great work, including their annual Star Count. Here is where you find the details on how to take part each February.
https://www.cpre.org.uk/what-we-care-about/nature-and-landscapes/dark-skies/

Mayo Dark Skies
If you're in Ireland, check in with Georgia and the Mayo Dark Sky Park. They keep their website updated with all upcoming events, in person and online.
https://www.mayodarkskypark.ie/about/our-story

Snowdonia Society
This society worked hard to help monitor the Dark Skies over Eryri in the early stages of the Dark Sky reserve bid. They run lots of volunteering days and are a great way to get out and about in the national park while giving something back to the area. You can become a member on their website.
https://www.snowdonia-society.org.uk/

Dark Source
Kerem Asfuroglu's wonderful lighting-design company. Looking for lighting design that puts community at its centre? You won't go wrong with Kerem.
https://www.dark-source.com/

Ridge and Partners
Andrew Bissell and his team are blazing a trail of darkness with their commercial-lighting schemes. They also work globally, so there's no excuse for not going dark.
https://ridge.co.uk/expertise/lighting-design/

Institute of Lighting Professionals
Find lots of great advice on lighting and find consultants or even lighting courses.
https://theilp.org.uk/

Buglife

Buglife are the awesome charity that works hard to stick up for our invertebrates. Find out how you can help save the small things that run the planet and support this organisation's excellent work here.

https://www.buglife.org.uk/

Resources/Useful links

These are just some of the organisations that work to save our wonderful creatures of the night and day. These are often charities, reliant on volunteers and donations.

I have used all of these organisations to research information for these creatures, and I highly recommend exploring the great work they all do.

Underneath every creature you will notice its 'IUCN Listing'. The International Union for Conservation of Nature's Red List of Threatened Species (IUCN) is the most 'comprehensive list of the global conservation status of animal, fungi and plant species'. It is an incredibly important tool that everyone in conservation uses to keep a check on the health of species world-wide, using it to influence change from the heady heights of the political sphere, to those who work at the frontline to keep our natural world alive.

Global

Dark Sky International
https://darksky.org/
Friends of the Earth
https://friendsoftheearth.uk
Sea Shepherd
https://seashepherd.org/
The Nature Conservancy
https://www.nature.org/en-us/

International Fund for Animal Welfare
https://www.ifaw.org
The International Union for Conservation of Nature
https://www.iucnredlist.org/
Worldwide Wildlife Fund
https://www.worldwildlife.org
Ocearch
https://www.ocearch.org/tracker/

Africa

African Pangolin Working Group
https://africanpangolin.org
African Wildlife Foundation
https://www.awf.org/
The Black Mambas
https://helpingrhinos.org/black-mambas
Tusk
https://tusk.org

Asia

Chengdu Research Base of Giant Panda Breeding
https://www.panda.org.cn/en/
Japan Tropical Forest Action Network
https://en.jatan.org
The Nature Conservation Society of Japan
https://www.nacsj.or.jp/english/

Oceania

Australian Wildlife Conservancy
https://www.australianwildlife.org/
Australian Space Agency
https://www.space.gov.au/

Bush Heritage Australia
https://www.bushheritage.org.au/
Environmental Defenders Office
https://www.edo.org.au/
CAFNEC
https://cafnec.org.au/
The Indigenous Rangers Program
https://www.niaa.gov.au/our-work/environment-and
-land/indigenous-rangers-program-irp
Hawai'i Conservation Alliance
https://www.hawaiiconservation.org
Kākāpō Recovery
https://www.doc.govt.nz/our-work/kakapo
-recovery/
New Zealand Department of Conservation
https://www.doc.govt.nz/
New Zealand Nature Fund
HYPERLINK https://nznaturefund.org

Papa New Guinea

Mahonia Na Dari
HYPERLINK https://www.mndpng.org
WCS Papua New Guinea
https://png.wcs.org/

North America

Cornell Lab of Ornithology
https://www.birds.cornell.edu/
National Audubon Society
https://www.audubon.org/
National Parks Conservation Association
https://www.npca.org

South America

Amazon Conservation
https://www.amazonconservation.org
Galapagos Conservation Trust
https://galapagosconservation.org.uk
Institute of Biology of the National Autonomous University of Mexico
https://www.restauracionecologica.org
Neotropical Primate Conservation
https://neoprimate.org

Europe

Bat Conservation Trust
https://www.bats.org.uk
British Divers Marine Life Rescue
https://bdmlr.org.uk
Buglife
https://www.buglife.org.uk
Butterfly Conservation
https://butterfly-conservation.org/
Royal Society for the Protection of Birds
https://www.rspb.org.uk
The Barn Owl Trust
https://www.barnowltrust.org.uk
The Mammal Society
https://www.mammal.org.uk/
The Zoological Society of London
https://www.zsl.org/
The Wildlife Trusts
https://www.wildlifetrusts.org

Acknowledgements

I would like to thank Ryan, for being my biggest believer, sanity checker and dream supporter. Even when this book seemed an impossible feat you stood strong and dragged me to the finish line. To my dad, Robert, for passing on to me the curiosity gene and my mum, Janet, for always being proud of me.

With thanks to Samara, Joshua, Tamasin, Zachary, Oliver-James, Xander, Hallie and Kuzey.

The Phillips crew; Scotty, Caz, Auntie Siân, Lauren and Cam.

My biggest cheerleaders; Charlotte, Rich and Maisie Lewis, Sarah 'Sally-Anne' Tracey, Kat Lawrenson, Kirk 'but actually Kirsty' Paton, Hanna Elin, Eleanor B and Gem Simmons.

Special thanks to Ben McConnell for working with me again and bringing Creatures to the real world.

Meg, Alice, Taslima and the rest of the Harper North team - thank you for your answers to my endless queries!

Team Eryri National Park; thank you for allowing my me have the best job in the world and being fantastic colleagues, especially team Cadwraeth.

Finally, the biggest thanks of all goes to conservationists around the world, who are working tirelessly to put a permanent halt to the decline of all these Creatures and many more. Conservation is a career that will bring you immense joys, crushing lows, lift your spirits and break your heart. It's not for the faint hearted. To each and every one of you working in the field to keep these creatures from slipping between our fingers into the pages of the history books; I thank you, on behalf of the wildlife you fight for every day, and for each and every person on the planet. Especially the ones who won't know what we are missing until it's gone. Don't give up, no matter how impossible it seems. To say we tried and gave it our all, even if we still lose, is the most noble of lives to live; to give a voice to the voiceless, the forgotten, and the barely existing.

Harper
North

Book Credits

HarperNorth would like to thank the following staff and
contributors for their involvement in making this book a
reality:

Fionnuala Barrett
Peter Borcsok
Laura Braggs
Sarah Burke
Cherie Chapman
Alan Cracknell
Jonathan de Peyer
Anna Derkacz
Tom Dunstan
Kate Elton
Sarah Emsley
Simon Gerratt
Imogen Gordon Clark
Lydia Grainge
Monica Green
Natassa Hadjinicolaou
Emma Hatlen
Jess Haycox
Megan Jones

Jean-Marie Kelly
Ben McConnell
Taslima Khatun
Holly Kyte
Rachel McCarron
Millie Morton
Alice Murphy-Pyle
Adam Murray
Genevieve Pegg
Amanda Percival
Dean Russell
Florence Shepherd
Colleen Simpson
Eleanor Slater
Hilary Stein
Emma Sullivan
Katrina Troy
Claire Ward
Ben Wright

For more unmissable reads,
sign up to the HarperNorth newsletter at
www.harpernorth.co.uk

or find us on X at
@HarperNorthUK

Harper
North